Tissue Engineered Scaffolds for Nerve Regeneration

With Best Regards
to Dr. Bellamkonda,
- Mahesh

Mahesh Chandra Dodla

Tissue Engineered Scaffolds for Nerve Regeneration

Gradients of Molecules Enhance Nerve Regeneration

VDM Verlag Dr. Müller

Imprint

Bibliographic information by the German National Library: The German National Library lists this publication at the German National Bibliography; detailed bibliographic information is available on the Internet at
http://dnb.d-nb.de.

Cover image: www.purestockx.com

Publisher:
VDM Verlag Dr. Müller Aktiengesellschaft & Co. KG, Dudweiler Landstr. 125 a,
66123 Saarbrücken, Germany,
Phone +49 681 9100-698, Fax +49 681 9100-988,
Email: info@vdm-verlag.de
Zugl.: Atlanta, Georgia Institute of Technology, 2000.

Produced in Germany by:
Schaltungsdienst Lange o.H.G., Zehrensdorfer Str. 11, 12277 Berlin, Germany
Books on Demand GmbH, Gutenbergring 53, 22848 Norderstedt, Germany

Impressum

Bibliografische Information der Deutschen Nationalbibliothek: Die Deutsche Nationalbibliothek verzeichnet diese Publikation in der Deutschen Nationalbibliografie; detaillierte bibliografische Daten sind im Internet über http://dnb.d-nb.de abrufbar.

Coverbild: www.purestockx.com

Verlag: VDM Verlag Dr. Müller Aktiengesellschaft & Co. KG, Dudweiler Landstr. 125 a,
D- 66123 Saarbrücken,
Telefon +49 681 9100-698, Telefax +49 681 9100-988,
Email: info@vdm-verlag.de
Zugl.: Atlanta, Georgia Institute of Technology, 2000.

Herstellung in Deutschland:
Schaltungsdienst Lange o.H.G., Zehrensdorfer Str. 11, D-12277 Berlin
Books on Demand GmbH, Gutenbergring 53, D-22848 Norderstedt

ISBN: 978-3-8364-5763-7

ACKNOWLEDGEMENTS

This work could not have been completed without the guidance, collaboration, and friendship of a number of people. I would like to take this opportunity to express my sincere thanks to them for their contributions.

I would first like to express my deepest appreciation to my PhD. thesis advisor, Prof. Ravi V. Bellamkonda, for his support, encouragement and guidance. I would like to thank all the members, both past and present of the Neurological Biomaterials and Therapeutics laboratory at Georgia Institute of Technology, Atlanta, U.S.A. I would like to acknowledge the financial support provided by a grant from the National Institute of Health (1R01 NS44409 to Prof. Ravi V. Bellamkonda), and a grant from Georgia Tech/Emory Center (GTEC, NSF EEC-9731643 to Prof. Ravi V. Bellamkonda).

TABLE OF CONTENTS

LIST OF TABLES

LIST OF FIGURES

SUMMARY

Nerve autografts are widely used clinically to repair injured nerves. However, nerve grafts have many limitations, including availability and loss of function at donor site. To overcome these problems, we have used a tissue engineering approach to design three-dimensional (3D) scaffolds containing laminin-1 (LN-1) and nerve growth factor (NGF). After nerve injury and during embryonic development, several extra-cellular matrix molecules and neurotrophic factors contribute to nerve regeneration and repair. Our 3D agarose hydrogel scaffolds with gradients of LN-1 and NGF are designed to mimic these *in vivo* conditions to promote nerve regeneration in rats.

To determine the effect of LN-1 gradients on neurite extension *in vitro*, dorsal root ganglia (DRG) from chick embryos were cultured in 3D hydrogels. A gradient of LN-1 molecules in agarose gels was made by diffusion technique. LN-1 was then immobilized to the agarose hydrogels using a photo-crosslinker, Sulfo-SANPAH (Sulfosuccinimidyl-6-[4'-azido-2'-nitrophenylamino] hexanoate). Anisotropic scaffolds with three different LN-1 gradients were used. Isotropic scaffolds with different uniform concentrations of LN-1, were used as positive controls. DRG cultured in anisotropic scaffolds with optimal LN-1 gradients extended neurites twice as fast as DRG in optimal concentration in isotropic scaffolds. Furthermore, in the anisotropic scaffolds the faster growing neurites were aligned along the direction of LN-1 gradient.

To promote nerve regeneration *in vivo*, tubular polysulfone guidance channels containing agarose hydrogels with gradients of LN-1 and NGF (anisotropic scaffolds) were used to bridge 20-mm nerve gaps in rats. Nerve autografts were used as positive controls and isotropic scaffolds, with uniform concentration of LN-1 and NGF, were used as negative controls. After 4-months, the rats were sacrificed and nerve histology analysis was done to test for nerve regeneration. Only anisotropic scaffolds and nerve autografts contained evidence of axonal regeneration. Both groups had similar numbers of myelinated axons and similar axonal-diameter distribution. However, the nerve grafts performed better in functional outcome as measured by relative gastrocnemius muscle weight (RGMW) and electrophysiology analyses. Optimization of performance of anisotropic scaffolds by varying the LN-1 and NGF concentration gradients might lead to development of scaffolds that can perform as well as nerve auotgrafts for nerve regeneration over long nerve gaps.

Chapter 1. **INTRODUCTION**

1.1. Aim of Thesis

The aim of my thesis is to develop *anisotropic* tissue-engineered scaffolds for promoting enhanced nerve regeneration across a challenging 20-mm long nerve gap in rats, as compared to *isotropic* scaffolds.

1.2. Problems and Challenges of Peripheral Nerve Injuries

Injuries to the peripheral nervous system (PNS) occur frequently and are a major source of disabilities. PNS injuries impair the ability to move muscles, to feel normal sensations, and result in painful neuropathies. As neurons are terminally differentiated, repair of injuries to the nervous system necessitates regeneration of neural tissue.

PNS injuries are classified as being *traumatic, non-traumatic or surgical* in nature. *Traumatic* nerve injuries result from collisions, motor vehicle accidents, gunshot wounds, fractures, lacerations or other forms of penetrating trauma. In 2002, more than 250,000 patients suffered *traumatic* peripheral nerve injuries in the U.S.A [56]. Even among the patients who received treatment, more than 50% did not show any measurable signs of recovery or suffered from drastically reduced muscle strength. Most of the *non-traumatic* peripheral nerve injuries are attributed to nerve compression and adhesion. In 2002, more than 400,000 repair procedures were done in U.S.A. to correct carpal tunnel syndrome, a *non-traumatic* nerve injury. Although treatments for *non-traumatic* nerve injuries have higher efficacy than *traumatic* nerve injuries, patients still suffer from pain, loss of muscle strength and dexterity for several weeks to months. *Surgical* injuries result from procedures such as prostatectomy, to remove prostate tumors. Prostatectomy procedures most often require sacrificing one or both of the cavernous nerves, adversely affecting erectile function and bladder control. In 2002, more than 260,000 patients in U.S.A. suffered major injuries to cavernosal nerves due to prostatectomy procedures [56]. In order to repair these nerve injuries, several techniques have been used.

The simplest technique for nerve repair, in case of nerve transection injuries, is reconnecting the two ends of the nerve using sutures or fibrin glue. However, often there is a loss of nerve segment due to injury, or there might be a time lag between injury and surgical repair during which the nerve ends might retract, resulting in a nerve gap. In such cases, end-to-end nerve suturing cannot be done without creating tension in the nerve segment, resulting in a poor regeneration outcome [115]. To overcome this problem, the two nerve ends are approximated using grafts, such as nerve autografts/allografts, muscle grafts, vein grafts, muscle-vein grafts and synthetic nerve guidance conduits/channels (NGCs).

Historically, during the 19th and early-20th centuries, various materials were used to promote nerve repair, such as bone [44], metal tubes [97], blood vessels [129] and fat sheaths [67], to guide the regenerating axons to the distal stump. The use of autologous nerve grafts was also first reported during this time [5]. However, due to improper surgical techniques, anatomical repair rarely led to an appreciable return of function. During World War II, surges in numbers of nerve injuries led to advances in microsurgical techniques and instrumentation. Further refinements in microsurgical techniques, revolutionized by Millesi [94], and drug therapies have had beneficial effects.

Significant advances in surgical techniques have been achieved, and now biological factors rather than surgical techniques are limiting improvements in nerve regeneration. Further advances may come from greater understanding of the molecular mechanisms of nerve regeneration, advances in nerve cell culture, development of new biomaterials and genetic techniques.

1.3. Nerve Autografts and Their Limitations

At present, nerve autografts are considered as the "gold standard" for bridging nerve gaps in cases of PNS injuries [83]. Autologous tissue grafts, such as nerve autografts, veins, muscles etc., possess several advantages. They are likely to be more biocompatible than artificial materials, are less toxic and provide a support structure to promote cell adhesion and migration [36]. However, there are several disadvantages with this technique. Obtaining natural graft leads to a loss of function, as well as potential neuroma formation at donor sites, requires multiple surgeries, and requires multiple small grafts are needed in case of long nerve gaps. There could be a problem of size mismatch between the donor nerve graft and injured nerve. In addition, complete functional recovery is seldom obtained even with autografts [70]. Therefore, there is interest in developing techniques to not only enhance the performance of autografts [34,47,136], but also synthesize alternatives with better functionality than autografts. Analytically, natural or artificial grafts used to bridge nerve gaps can be thought to have four central components germane to regeneration: 1) scaffold/substrate, 2) growth factors, 3) extracellular matrix (ECM) molecules and 4) cells. The current techniques for nerve regeneration try to provide all or combinations of the four components for nerve regeneration.

1.4. Polymer Nerve Guidance Channels for Nerve Regeneration

To replace nerve autografts for peripheral nerve regeneration, a wide variety of natural materials have been used, such as, bone, autologous muscle grafts [92], vein grafts [20,124], tendon grafts [14], muscle-vein combined grafts [96]. All have exhibited encouraging results in a research setting but suffer from the drawback that tissue must still be removed from the patient, leading to loss of other functions.

Tubular nerve guidance channels (NGC) made of synthetic materials has also been used to bridge nerve gaps with promising results [2,68,90]. NGCs with inner diameters slightly larger than the diameter of the nerve being repaired are used. This strategy involves securing the proximal and distal ends of the nerve to the two open ends of the NGC. These tubes physically guide or funnel sprouting axonal processes towards their distal targets. In addition to restricting migrating cells and sprouting processes within the confines of the tube, such systems create a space for increasing the local growth factor concentrations and act as a barrier to minimize the infiltration of fibroblasts and other inhibitory or scar forming entities that would otherwise inhibit the normal regenerative process [120]. Guidance channels have been widely used in tissue engineering to repair peripheral nerve injuries; the various kinds of NGC being used are discussed in more detail in section 1.8.2.

1.5. Nerve Regeneration using NGC

Nerve regeneration in silicone NGC has been studied in detail [130]. Within a few hours of nerve repair using NGC, the tube fills with serum exuded by the injured blood vessels in the nerve ends. This fluid contains neurotrophic factors, as well as several cytokines and inflammatory cells such as macrophages. The macrophages help remove the myelin and axonal debris formed due to injury. The fluid also contains the clot-forming protein, fibrin. From day 3 to day 7 post-repair, fibrin coalesces and forms a longitudinally oriented fibrin cable bridging the two nerve ends. Without the formation of the fibrin cable, axonal regeneration cannot occur, thus making the fibrin cable formation a very critical step. From day 7 to day 14, the fibrin cable is invaded by cells migrating from the proximal and distal nerve stumps, including fibroblasts, macrophages, Schwann cells and endothelial cells (which form capillaries and larger vessels), followed by axons. From 2 to 8 weeks, some of the axons from the proximal end grow into the fibrin matrix, reach the distal nerve end and proceed to the final target tissues, and are simultaneously myelinated by Schwann cells. During nerve regeneration, axons are guided to their targets by a combination of contact mediated cues such as extracellular matrix (ECM) proteins, and diffusible cues including neurotrophic factors secreted by Schwann cells and macrophages [40,116].

1.6. Role of Schwann Cells in Nerve Regeneration

In the PNS, Schwann cells are support cells that wrap around the axons. A Schwann cell forms a multilamellar sheath of myelin, a phospholipid-containing substance, around an axon that serves as an insulator and increases nerve conduction velocity. An individual Schwann cell may ensheath several unmyelinated axons, but only one myelinated axon, within its cytoplasm.

Schwann cells of uninjured nerves are quiescent. Following nerve injury, the Schwann cells become 'reactive' and proliferate. Schwann cells migrate into the injury site and then form cell cords (known as bands of Büngner) within the basal lamina tubes, and the regenerating axons grow in close contact with these cords [16]. Impeding the migration of Schwann cells prevents the growth of injured axons through the grafts [53].

The 'reactive' Schwann cells have been shown to offer a highly preferred substrate for axon migration and release bioactive factors that further enhance nerve migration [7]. For instance, Schwann cells produce ECM molecules such as laminin and collagen, and express many cell adhesion molecules and receptors including L1, N-cadherin, γ1 integrins, and neural cell adhesion molecule (N-CAM). Schwann cells also synthesize and secrete a cocktail of neurotrop(h)ic molecules such as nerve growth factor (NGF), brain derived neurotrophic factor (BDNF), and ciliary neurotrophic factor (CNTF) [117]. These ECM proteins and neurotrophic factors are essential for the survival of neurons and axonal migration [26,39,54,87]. Researchers have attempted to take advantage of the cellular functions of Schwann cells to promote regeneration in both PNS and CNS, as described below.

1.6.1. Techniques Using Schwann Cells for Nerve Regeneration

Using Schwann cells in NGCs bypasses the fibrin cable formation step, accelerates the formation of bands of Büngner, and introduces a persistent source of neurotrophic factors, leading to more efficient nerve repair. When saline filled guidance channels are implanted, Schwann cell migration into their lumen is a critical step for peripheral nerve regeneration [37,53]. Using Schwann cells has lead to enhanced nerve regeneration in both PNS and CNS [48,131]. When Schwann cells are seeded into nerve conduits, the injured peripheral nerves regenerate at a much faster rate [15,51]. This could decrease the time required by the axons to reconnect to the target organ, as well as increase the distance over which regeneration occurs. Addition of Schwann cells has been shown to significantly improve the performance of various scaffolds, such as empty NGCs, collagen gels, venous nerve grafts and muscle grafts, as compared to control scaffolds without Schwann cells [7,64,112]. Therefore, Schwann cells isolated from the peripheral nerve of a patient, and expanded *in vitro*, could be used to treat the patient's nerve injuries.

The success of nerve auotgrafts for peripheral nerve regeneration is attributed primarily to the Schwann cells they contain. For instance, acellular nerve segments do not function as well as Schwann cell-containing autografts [125,126]. The ability of Schwann cell-seeded NGCs to promote regeneration was found to be dependent on the Schwann cell seeding density and immunocompatibilty between donors and host [48]. Using Schwann cells from a genetically identical source (syngenic) in rats, it was observed that increasing the seeding density improves the nerve regeneration outcome. Heterologous Schwann cells elicited a strong immune reaction, impeding the nerve regeneration [48]. Also, the ability of NGCs containing Schwann cells to facilitate nerve regeneration is dependent on the density of pre-seeded Schwann cells [48]. To obtain Schwann cells from a patient, an additional surgery is needed to obtain a small nerve segment that will serve as source of Schwann cells, and it takes 5-6 weeks to grow a therapeutically relevant number of Schwann cells, during which time the injured nerve further degenerates [79]. Therefore, the current techniques for Schwann cell transplantation are not clinically acceptable.

1.6.2. Techniques Using Other Cell Sources for Nerve Regeneration

As an alternative to Schwann cells, other cells could be used as such or genetically modified, to produce desired levels of neurotrophic factors, or to express specific extracellular matrix molecules. Fibroblasts, genetically modified to produce NGF, BDNF, NT-3 and bFGF, showed promising results in CNS regeneration [95]. Olfactory ensheathing cells have been shown to promote regeneration of cut nerves in the adult rat spinal cord [80]. Although these are examples of CNS regeneration, the genetically modified cells can be used for PNS regeneration also. Addition of bone marrow stromal cells to NGCs has shown improved regeneration over empty NGCs in a rabbit peroneal nerve model [23]. Similarly, pluripotent stem cells derived from hair follicles have shown improvements in mice [6]. However, the difficulties of isolating and culturing these cells from the patient prior to surgery, could limit this approach for clinical applications.

1.7. ECM Proteins, Neurotrophic Factors and PNS Regeneration

Schwann cells have been shown to offer a highly preferred substrate for axonal migration and release bioactive factors that further enhance this process. Schwann cells produce two kinds of molecules that are relevant for nerve regeneration: *haptotactic* proteins, that are substrate-bound or cell surface-bound extracellular proteins, such as laminin-1 (LN-1), collagen, L1, N-cadherin, and NCAM; and diffusible *chemotactic* proteins such as nerve growth factor (NGF), brain derived neurotrophic factor (BDNF), and ciliary neurotrophic factor (CNTF). The success of peripheral nerve and Schwann cell transplants is essentially due to their ability to secrete ECM proteins such as LN-1 and trophic factors like NGF in response to injury [33,81].

1.7.1. ECM Protein, Laminin

The ECM molecules are present in the basement membrane of cells, and include laminin, fibronectin, collagen, tenascin, thrombospondin and heparan sulfate proteoglycans. ECM molecules, especially laminin and fibronectin, play vital roles in cell migration, adhesion, proliferation, differentiation and neurite outgrowth in the nervous system during development and regeneration [77,99].

Laminin is the first ECM molecule to be expressed during embryonic development, suggesting its high importance during ECM organization [82]. Laminin is a potent stimulator of neurite outgrowth, enhances attachment and migration of neuronal cells, increases survival of neurons and stimulates neurite outgrowth in neurons from peripheral and central nervous systems [1,9,13,25,29,101,102]. After peripheral nerve injury native Schwann cells have been shown to increase production of laminin to induce regeneration. When nerve grafts pre-treated with laminin-1 antibodies were transplanted across a sciatic nerve gap in rodents, a 52% decrease in axon ingrowth into nerve grafts was observed [125], suggesting that laminin-1 is critical for nerve regeneration.

Laminin is a large cross-shaped glycoprotein (M_w = 900,000) found in the basement membrane of most non-neural tissues and dispersed through out the nervous system. The Laminin molecule is a heterotrimeric molecule, composed of three polypeptide chains: α, β and γ. These chains are structurally homologous but differ in their amino acid sequences, resulting in different isoforms (i.e. $\alpha1$, $\alpha2$, $\alpha3$ etc.) and hence different members of the laminin family. Laminin has at least 10 isomers, from laminin-1 to laminin-10 [82].

Localization studies using isoform specific antibodies in the human fetal tissues (18-19 weeks) have revealed that laminins are expressed in a tissue specific manner [82]. Laminin-1 ($\alpha1$-$\beta1$-$\gamma1$) is expressed in the brain, especially in the neuroretina, the olfactory bulb, and cerebellum, meninges, as well as in kidney and testis. Laminin-1 was the first member of the laminin family to be isolated and is also the best studied member. Laminin-1 can be readily purified in large quantities from Engelbreth-Holm-Swarm (EHS) tumor. Hence, most studies of the effects of laminin *in vitro* have been carried out with laminin-1. Laminin-1 promotes neurite formation, neurite outgrowth and neuronal survival. In addition, laminin-1 guides axonal path finding, cell migration and adhesion [82]. Laminin-2 ($\alpha2$-$\beta1$-$\gamma1$) is expressed in cardiac and skeletal muscle, and peripheral nerve. Laminin-2 promotes neurite extension and mediates Schwann cell myelination [82]. Laminin-3 ($\alpha1$-$\beta2$-$\gamma1$) is expressed in neuromuscular junctions and renal glomeruli (kidneys). Laminin-3 helps in recognition of synaptic sites during development and regeneration. Laminin-4 ($\alpha2$-$\beta2$-$\gamma1$) is expressed in Schwann cells and skeletal muscle. Laminin-4 is involved in Schwann cell myelination. Laminin-5 ($\alpha3$-$\beta3$-$\gamma2$) and laminin-6 ($\alpha3$-$\beta1$-$\gamma1$) are expressed primarily at the dermo-epidermal junctions. Laminin-7 ($\alpha3$-$\beta2$-$\gamma1$) is expressed in the human amnion. Laminin-8 ($\alpha4$-$\beta1$-$\gamma1$) and -9 ($\alpha4$-$\beta2$-$\gamma1$) are expressed primarily in the human chorion. Laminin-10 (α5-β1-γ1) is expressed in the human amnion and keratinocyte cultures.

On the laminin molecule, two specific segments of peptide sequences (IKVAV and LQVQLSIR) have been identified for promoting neurite outgrowth and cell attachment [82]. However, these sites are not active with all laminin-responsive neuronal cells, and some cell-type specificity is observed in the neuronal cell response. This suggests that additional sites on laminin exist for neurite outgrowth and that more than one receptor for signaling neurite growth may exist.

Cells interact with laminins through a number of cell surface receptors and binding proteins. The receptors and binding proteins that bind laminin to cell surfaces and mediate adhesion and neurite outgrowth are: (1) Integrins of the β family; (2) Non-integrin binding proteins that bind specific sequences in one of the polypeptide chains; and (3) Carbohydrate-binding moieties such as lectins that bind galactose and pol-N-acetyllactosamine groups in oligosaccharides. The binding of laminin to its receptors activates and initiates the signals mediated by protein kinase cascades, GTPase, and intracellular calcium fluxes. These secondary signals modulate a variety of cell responses such as attachment to a substrate, migration and neurite outgrowth during embryonic development and regeneration [69].

1.7.2. Neurotrophic factor, NGF

The distal end of the transected nerve exerts trophic influences on regenerating axons [17]. Neurotrophic factors, including neurotrophins such as NGF and BDNF are produced by Schwann cells following nerve injury and dispersed diffusely in a gradient fashion around regenerating axons. Regenerating axons extend along the density gradient of neurotrophins to the distal segment [72]. Apart from NGF and BDNF, Schwann cells also produce other neurotrophins, such as, insulin-like growth factor (IGF), ciliary neurotrophic factor (CNTF) and fibroblast growth factor (FGF). Neurotrophic factors are crucial for the development and survival of the mammalian nervous system [78].

Although all the above compounds stimulate peripheral nerve migration directly or indirectly, NGF is a very important and best-characterized neurotrophic factor for peripheral nerve regeneration. NGF is a dimeric, 27 kDa neurotrophic factor. NGF is produced in the target organs of the sympathetic and sensory nerves, and is conveyed to the cell body retrogradely in the PNS [8]. Intact sympathetic and sensory nerves contain low titer of NGF-mRNA as compared to their target organs [54]. However, following axotomy, Schwann cells in the distal segment start producing NGF and reach a peak at 24 h. This 10- to 15-fold elevated level is maintained for at least 2 weeks after axotomy [54]. NGF has been shown to stimulate and promote the survival of sensory ganglia and nerves, including spinal sensory nerves and sciatic nerves [78,118]. Griffin et al. [46] have shown concentration-dependent neurite extension/retraction with NGF levels. NGF has been shown to prevent the death of axotomized sensory neurons completely following exogenous administration [100]. There is enough evidence to show that gradients of neurotrophins help in guiding developing as well as regenerating axons in a wide variety of neuronal systems [49,50].

Receptors for NGF include trkA (a NGF-specific, high affinity receptor, 40 kDa glycoprotein) and $p75^{NTR}$ (a low affinity, 75 kDa glycoprotein) [98]. NGF receptors, after binding to NGF, become activated by the ligand mediated oligomerization of receptors and the following autophosphorylation of their tyrosine residues [61]. The activation of NGF receptors initiates a cascade of signaling events and leads to neuronal differentiation [107].

1.7.3. The Combined effects of ECM proteins and Neurotrophic Factors for PNS Regeneration

In neuronal development and in adult nervous systems, substrate-bound neurite promoting ECM proteins and chemoattractive diffusible trophic factors influence and stimulate axon guidance and neurtite extension [66,99]. ECM proteins affect cell interactions during the development, maintenance and regeneration of the nervous system [58]. Neurotrophic factors support growth, differentiation and survival of neurons in the nervous system [65].

LN-1 and NGF have been used separately in different studies to promote neuronal survival and extension. However, LN-1 and NGF, used together, promoted significantly better nerve regeneration than either one of them used separately [132], suggesting that LN-1 and NGF might be acting synergistically to promote better neurite extension.

1.8. Tissue Engineering Strategies for PNS Regeneration

Tissue engineering techniques are being used to develop constructs to replace natural nerve grafts. Tissue engineered constructs for nerve regeneration consist of one or more of the four components: (i) scaffold, usually tubular NGC along with structural elements such as fibers, gels etc., (ii) growth factors, (iii) ECM proteins, and (iv) cells. In general, scaffolds for nerve repair should support axonal proliferation, have low antigenicity, support vascularization, be porous for oxygen diffusion and avoid long term compression. The scaffold can be made from natural or synthetic materials.

1.8.1. Natural Materials as Scaffolds

Natural materials, including veins, skeletal muscle fibers and collagen, have been used to make scaffolds. Autologous vein grafts have been shown to provide a good environment for axonal regeneration in short nerve gaps [38,127]. However, use of vein grafts for long nerve gaps has been less successful, because of the collapse of thin-walled veins, and constriction due to the surrounding scar tissue [22]. In order to prevent vein grafts from collapsing and improve their performance, intraluminal space fillers such as autologous Schwann cells, collagen, and muscle fibers have been used. Collagen-filled vein grafts were found to promote better axonal growth than empty vein grafts for a 15-mm nerve gap in rabbits [24]. Similarly, Schwann cell-seeded venous grafts supported axonal growth and performed better than unseeded grafts to repair 40 mm nerve gaps [134] and 60 mm nerve gaps in rabbits [112]. The principal drawback of this approach is that it requires the availability of relevant amount of live autologous Schwann cells (up to 8 million cells/ml) that is difficult to obtain. Muscle-vein combined grafts, in which the muscle fibers are inserted in veins, were used in 10-mm long nerve defects in rats, and found to promote axonal regeneration comparable to that of syngenic nerve grafts [41]. Although the muscle-vein grafts were able to promote nerve regeneration in 55-mm long nerve defects in rabbits, they were not comparable to nerve autografts [41]. Autologous muscle-vein combined grafts have been used clinically in humans to bridge nerve gaps ranging from 5-60 mm. The results were scored as “poor”, “satisfactory”, “good” and “very good”, based on recovery of sensory and motor functions. Of the 21 lesions repaired (in 20 patients), 10 were lesions of the sensory nerves and 11 were mixed nerve lesions. All lesions in the sensory nerves, except one greater than 30 mm, showed “good” to “very good” recovery. All lesions in the mixed nerves showed “satisfactory” to “good” recovery of motor and sensory functions [10].

Although autogenous/natural materials have shown encouraging results when used for nerve repair, they still have certain drawbacks. In case of autogenous grafts, the drawbacks include the need for a second surgery, the loss of function and neuropathic pain at donor site. Allografts have problems related to preservation and immuno-rejection. In order to avoid these problems, grafts made of artificial/synthetic materials have been used.

1.8.2. Synthetic Scaffolds for Nerve Repair

To overcome the problems associated with harvesting natural tissues, synthetic scaffolds have been developed. Among the artificial materials, synthetic tubular NGCs have shown the most promising results so far (Figure 1-1). The use of NGC reduces tension at the suture line, protects the regenerating axons from the infiltrating scar tissue, and directs the sprouting axons toward their distal targets. The luminal space of NGCs can be filled with growth promoting matrix, growth factors and/or appropriate cells. In some cases of nerve repair, NGCs have been used to intentionally leave a small gap between the injured nerve ends, to allow accumulation of cytokines, growth factors and cells [27]. The NGCs can be used as an excellent experimental tool, to control the

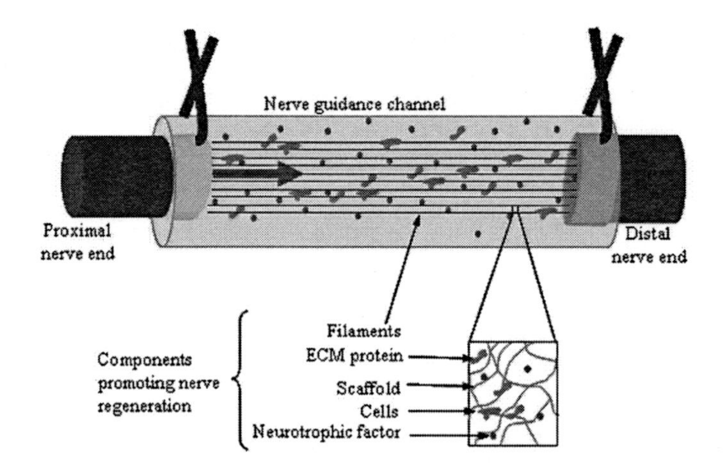

Figure 1-1: A schematic of a synthetic nerve guidance channel (NGC). The NGC, sutured to the nerve ends, is filled with one or more of the following components: hydrogel, filaments, cells, neurotrophic factors and ECM proteins. For an *isotropic* graft, there would be no filaments and the other components would be distributed uniformly. For an *anisotropic* graft, there may be filaments, and the other components would be aligned longitudinally or in increasing concentration from proximal to distal nerve end.

distance between the nerve stumps, test the fluid and tissue entering the channel, and vary the properties of the channel [122].

Nerve regeneration in silicone NGCs has been studied in detail [130], as described in section 1.4. Briefly, following nerve repair using a NGC, a fibrin cable connecting the two nerve ends is formed within the NGC. The fibrin cable is then invaded by cells migrating from the proximal and distal nerve stumps. Axons from the proximal end grow into the fibrin matrix and are engulfed by the Schwann cells. Some of these axons then reach the distal nerve end and get myelinated. In inert silicone tubes of 10 mm or shorter, these processes occur spontaneously. However, it is generally accepted that impermeable, inert NGCs such as silicone do not support regeneration across defects larger than 10 mm, without the presence of exogenous growth factors. The regeneration process can be improved by various approaches like changing properties of the tube (permeability, roughness and electric charge characteristics), addition of matrices, neurotrophic factors, ECM molecules and cells [122].

Based on the porosity and/or degradability of the material used, NGCs can be classified as impermeable, semi-permeable and resorbable (Table 1-1).

Table 1-1: Classification of NGCs based on porosity and degradability

Porosity	Degradability	Example [reference]
1. Impermeable	Non-degradable	Silicone [48]
2. Semipermeable	Non-degradable	PS [80], PAN/PVC [1, 71]
3. Resorbable	Degradable	PLA [10], PGA

Porosity affects the movement of soluble factors, oxygen and waste products, through the walls of the NGCs, which is vital for nerve regeneration. Silicone tube is an example of an impermeable NGC since it does not permit movement of molecules across the tube walls. Examples of semi-permeable tubes are polysulphone (PS) and polyacrylonitrile/polyvinylchloride (PAN/PVC). Nerves regenerated in semipermeable tubes featured more myelinated axons and less connective tissue, as compared to impermeable NGCs [2,121]. PAN/PVC channels with a molecular weight cutoff of 50 kilodaltons support regeneration even in the absence of a distal nerve stump [2]. Examples of bioresorbable tubes are, polylactic acid (PLA), polyglycolic acid (PGA), poly(L-lactide-co-glycolide) (PLGA), poly-(lactide-co-caprolactone) (PLC) and poly(3-hydroxybutyrate) (PHB). The use of bioresorbable tubes negates the need for a second surgery to remove the implant and prevents long-term compression of the nerve. However, it is critical that the degradation of the tube should not allow fibroblasts to invade the lumen space before regeneration occurs, as this may prevent axons from regenerating.

A wide variety of polymeric materials (polyethylene, silicone elastomers, polyacrylonitrile etc.), metals (stainless steel, tantalum etc.) and biological materials (artery, collagen, vein etc.) have been used to make these guidance channels [21,42,85,113]. In our study we have used polymeric guidance channels made of polysulfone (PS). PS is a semi-permeable material having a molecular weight cut-off of 50 kDa, which allows for controlled solute exchange between the internal regenerative and the external wound-healing environment, preventing connective tissue formation [2,121].

1.8.3. Polymer Guidance Channels Containing 3D Hydrogel Scaffolds

Collagen gels

Collagen gels and filaments have been used to promote PNS regeneration. Collagen gel can be used to fill the intra-luminal space of a vein graft to prevent it from collapsing and improve its nerve repair efficiency. In collagen-filled vein grafts, the number and diameter of myelinated axons was significantly increased compared to vein grafts without collagen gel [24]. Nerve repair with silicone tubes can be significantly improved by filling them with collagen gel. Collagen tubes filled with collagen gel have promoted more rapid nerve sprouting, and better morphology, than saline-filled collagen tubes [106]. However, in some cases collagen gels have hindered regeneration as compared to saline-filled tubes [123]. In addition, collagen gels may be immunogenic when implanted *in vivo* for long periods of time [119], which might in turn inhibit nerve regeneration.

Hyaluronic Acid

Hyaluronic acid, an ECM component, is associated with decreased scarring and improved fibrin matrix formation. It is hypothesized that during the fibrin matrix phase of regeneration, hyaluronic acid organizes the extracellular matrix into a hydrated open lattice, thereby facilitating migration of the regenerated axons [108]. Hyaluronan-based tubular conduits, used for peripheral nerve regeneration, resulted in more myelinated axons and higher nerve conduction velocities than silicone tubes filled with saline [128], with little cytotoxicity [60] upon degradation.

Other gels used *in vivo* to promote nerve regeneration include Matrigel, alginate gels, fibrin gels and heparin sulphate gels [31,86,114] with mixed results.

1.8.3.1. Agarose hydrogel is a potential biomaterial for bridging injured nerves

Agarose is a polysaccharide derived from red agar and is widely used in gel electrophoresis and gel chromatography. SeaPrep® agarose hydrogel has been shown to support neurite extension from a variety of neurons in a non-immunogenic manner [11,30,73]. In addition, SeaPrep® agarose is thermo-reversible, making it easy to work with. When heated to 65°C, SeaPrep® agarose hydrogels change to a liquid state and maintain this liquid state until cooled below 17°C, which results in the formation of gel. The gel-state is maintained until it is again heated above 65°C. SeaPrep® agarose is transparent, allowing cells cultured in it to be easily observed by light microscopy. Agarose gels also allow for molecules to be covalently linked to the gels through functional groups on its polysaccharide chains. For example, laminin protein or fragments of laminin can be covalently coupled to SeaPrep® agarose gels to enhance their ability to support neurite extension [133]. Although agarose gels support neurite growth on their own, coupling of molecules, such as laminin, significantly enhances the gels' ability to promote neurite extension. Due to these reasons agarose gels have been used to promote nerve regeneration, both *in vivo* and *in vitro*. 0.5% (w/v) SeaPrep® agarose gel was reported to promote higher functional recovery than 1% and 2% SeaPrep® agarose gels in sciatic nerve injury model in rats, and no immunological response was observed [73,132]. Hence we have used 0.5% SeaPrep® agarose gels for *in vivo* study. However, 1% agarose gel was used for *in vitro* study because agarose gel with higher strength was needed for *in vitro* work.

1.8.4. Polymer Guidance Channels with Neurotrophic Factors for Regeneration

Neurotrophic factors are produced by the target organs of nerves and help maintain the target organ-nerve synapse [59]. A nerve injury usually results in disruption of communication between the target organs and the neuronal cell body, and leads to Wallerian degeneration (breakdown of myelin sheath and axons). Due to cytokines released during Wallerian degeneration, Schwann cells are activated and produce neurotrophins such as nerve growth factor (NGF) and brain-derived neurotrophic factor (BDNF). Although many other trophic factors, including insulin-like growth factor (IGF), fibroblast growth factor (FGF) and ciliary neurotrophic factor (CNTF), have been shown to be involved in the promotion of nerve regeneration [43,45], it is believed that they are released from Schwann cells following mechanical damage to the cells.

NGF is produced in the target organs of sensory and sympathetic nerves in the PNS, and has been shown to stimulate and promote the survival of sensory ganglia and nerves, including spinal sensory nerves and sciatic nerves [78,118]. BDNF is expressed in very low levels in intact adult peripheral nerves, but is upregulated following injury. BDNF is effective in promoting the survival and outgrowth of not only sensory and sympathetic nerves, but also motor nerves [109].

Neurotrophic factors will likely be an important part of future clinical therapies for peripheral nerve injuries/diseases. Diseases in which the functions of Schwann cells are severely suppressed (Multiple Sclerosis, for example) or when acellular grafts (containing no viable Schwann cells) are used, delivery of neurotrophic factors could be highly effective in facilitating nerve regeneration. Various studies have utilized the functions of NGF to promote nerve regeneration. Hubble and Sakiyama-Elbert have developed a fibrin matrix that immobilizes heparin molecules by electrostatic interactions, which in turn immobilizes heparin binding growth factors. The fibrin matrix, when implanted *in vivo,* releases the bound growth factor due to fibrin degradation. This system was used to deliver NGF [76] for peripheral nerve regeneration *in vivo*, and bFGF [104] for neurite extension from chick dorsal root ganglia (DRG) *in vitro*. Fibrin-heparin-NGF matrix was observed to promote nerve regeneration comparable to syngenic nerve grafts over a 13-mm nerve gap in rats. Fibrin matrix that released bFGF enhanced neurite extension from DRG by 100% compared to unmodified fibrin matrix.

1.8.4.1. Lipid-based Microtubules for slow-release of NGF

NGF has been shown to facilitate nerve regeneration across long nerve gaps [28]. However, as neurotrophic factors have very short-half lives in vivo and are unstable in solution, controlled release systems are needed for their delivery [3,32]. For example, basic fibroblast growth factor released in a controlled manner significantly improved rat sciatic nerve regeneration across a 15-mm long gap [32]. In addition, slow release of growth factors can overcome the problem of "growth factor oasis" where neurites accumulate at high neurotrophic factor concentration regions in the scaffold.

Poly (ethylene-co-vinyl acetate) (EVAc) has been widely used for long-term release of NGF [55,71]. However, in this technique organic solvents are used for loading NGF into the EVAc polymer, and this loading process leads to dramatic loss of NGF activity [71]. To overcome this limitation, phosphatidylcholine based lipid microtubules have been developed for the delivery of bioactive agents [103,111]. Lipid microtubules are ideal for trophic factors because these sensitive proteins can be loaded into microtubules in an aqueous environment with no exposure to organic solvents. Also, because of their small dimension, the lipid microtubules can slowly release trophic factors without impeding neurite extension. The lipid microtubules can also be fabricated in different lengths to vary the release profile of the trophic factor.

1.8.5. Polymer Guidance Channels with ECM Molecules for Regeneration

Insoluble ECM molecules, such as laminin, fibronectin and certain forms of collagen, promote axonal extension and therefore, are excellent candidates for incorporation into the lumen of NGCs. Agarose gels crosslinked with laminin showed enhanced neurite extension from chick DRG *in vitro* [133]. Agarose gels crosslinked with laminin and soluble NGF showed nerve regeneration comparable to autografts over a 10 mm gap in rats [132]. However, axonal extension in the laminin gels depends on concentrations of laminin gels. High concentrations of laminin hinder regeneration [74]. Matrigel, a gel containing collagen type IV, laminin and glycosaminoglycans, supports some degree of regeneration over a long nerve gap in adult rats, when introduced into the lumen of NGCs [86]. Similarly, a gel mixture containing laminin, collagen and fibronectin, significantly improved nerve regeneration compared to saline-filled silicone NGCs [19].

1.8.6. Polymer Guidance Channels Pre-Seeded with Neuronal Support Cells for Nerve Regeneration

Performance of NGCs can be significantly improved by seeding the scaffolds with Schwann cells. Apart from Schwann cells, genetically modified fibroblasts, olfactory ensheathing cells, bone marrow stromal cells, pluripotent stem cells from hair follicles have been used to promote nerve regeneration [6,23,80,95]. These techniques have been discussed in more detail in section 1.6.1. - 1.6.2. However, the difficulties of isolating and culturing these cells from the patient prior to surgery, could limit this approach for some surgical procedures.

1.9. Challenges for the Design of Guidance Channels

An ideal guidance channel should be biodegradable, should not elicit immune response, have electrical activity and be porous. It should also be able to incorporate support cells, neurotrophic factors, and internal oriented matrices. Also, it should be flexible, readily available and easy to fabricate [57]. In designing a polymer nerve guidance channel, its characteristics, such as, porosity, roughness, and electrical activity, can be modified to enhance its properties.

Up to the present, many techniques have been examined for enhancing peripheral nerve regeneration using guidance channels. However, none of them have performed better than nerve autografts, the clinical "gold standard" for repairing nerve injuries. Even in cases where significant regeneration has been observed, it has been over short nerve gaps, about 10-mm nerve gaps. However, in clinical applications for humans, there is a need to bridge nerve gaps greater than 10-mm. Even in the best case of autografts, complete functional recovery is usually not achieved, due to misdirection of growing fibers, neuronal cell death, or atrophy/death of the denervated target organ [93,135]. Therefore, it is still a challenge to develop alternative approaches for enhancing peripheral nerve regeneration to the level comparable to autografts, and further to exceed the results obtained from autografts.

1.10. Need for 3D Culture Models and Anisotropic Scaffolds

1.10.1. Limitations of 2D tissue culture based models

Most previous studies have examined the effects of sharp boundaries of laminin itself or laminin with other molecules to direct neurite outgrowth [35,89] in 2D cultures. Even in other studies which have examined the directionality of axonal elongation on gradients of substrate-bound molecules, 2D cultures have been used [1,52,91]. However, *in vivo*, the axons have to navigate a 3D ECM environment, and the changes in the LN concentration are more likely to be gradual than sharp. The distribution of nutrients, substrate-bound molecules and cell-surface receptors in 3D cultures may be closer to *in vivo* conditions, as compared to 2D cultures. Hence 3D cultures may mimic *in vivo* conditions better than 2D cultures.

1.10.2. Limitations of Isotropic Scaffolds

Currently, tissue engineered scaffolds using neurostimulatory molecules to promote nerve regeneration are isotropic in nature, that is, these molecules are distributed uniformly throughout the scaffold. However, during development of nervous system, the molecules regulating neuronal migration (both permissive and inhibitory) are presented in a spatially and temporally controlled fashion [12,13,110].

In vitro studies have suggested that gradients of neurotrophic factors can direct growth cones towards the source of neurotrophic factor [49]. Insoluble and soluble gradients of NGF, NT-3 and BDNF have been shown to direct the growth of neurites from PC12 cells toward increasing concentrations of neurotrophic factors [18,63]. Therefore, *anisotropic* scaffolds having gradients of neurotrophic factors, along with other components might be an important tool for PNS regeneration.

Gels containing ECM molecules, such as laminin, collagen, fibronectin and glycosaminoglycans, have been widely used to make *isotropic* scaffolds for nerve regeneration. *In vitro* experiments with spatial patterns of whole ECM molecules [62] or their peptide derivates [105], have been used to direct the growth of neurites, as well as enhance neurite extension. *In vitro* experiments have demonstrated that gradients of ECM proteins could orient and enhance neurite outgrowth toward increasing concentrations of ECM molecules [1,29]. However, this technique is yet to be utilized to enhance nerve regeneration *in vivo,* due to difficulties in making gradients of proteins in three-dimensional scaffolds.

1.11. The Design and Significance of Our Study

The objective of this research is to develop agarose-based 3D scaffolds to bridge a nerve gap induced by injury, and promote the regeneration of the transected nerves across a challenging nerve gap of 20-mm in rats. The scaffolds have been incorporated with gradients of LN-1 and NGF to promote directional neurite extension of the regenerating axons. *The main hypothesis is:* since neurons use gradients of neuro-stimulatory and inhibitory cues for neurite extension and target finding during development and regeneration, such gradients will be suitable for stimulating peripheral nerve regeneration in adult animals. Hence, a spatially and temporally controlled presentation of neurite growth promoting molecules, NGF and LN-1, might lead to enhanced regeneration.

1.11.1. Design a novel *anisotropic* agarose hydrogel based 3D scaffolds to stimulate directional neurite extension *in vitro.*

In this study we designed an agarose hydrogel-based 3D scaffold with increasing concentrations of ECM protein, LN-1, to stimulate and direct the growth of neurite *in vitro*. The hypothesis is that *anisotropic* 3D agarose hydrogel scaffolds with gradients of LN-1 will promote longer and faster neurite extension than isotropic scaffolds with uniform concentrations of LN-1.

In the current work, we propose that a diffusion model can be used to create gradients of laminin molecules in porous agarose hydrogels. The LN-1 molecules can then be photo-immobilized within the agarose matrix to create immobilized gradients of LN-1. Isotropic LN-1 scaffolds, with uniform concentrations of LN-1 coupled to agarose hydrogel, have shown enhance neurite extension from primary chick dorsal root ganglia (DRG) and PC 12 cells [132]. In our current work, we would like to test if *anisotropic* scaffolds can promote longer and more oriented neurite extension, as compared to the *isotropic* scaffolds. Enhanced neurite extension using *anisotropic* scaffold will lead to design of better scaffolds for axonal regeneration *in vivo*.

1.11.2. Enhance nerve regeneration across a 20-mm long nerve gap in adult rats using *anisotropic* agarose-based scaffolds

In this study we propose the design of an agarose-based scaffold with increasing concentrations of two neurostimulatory molecules, LN-1 and NGF, to enhance nerve regeneration across a challenging nerve gap of 20-mm in rodents.

Polysulfone NGCs containing agarose hydrogel scaffolds will be used to bridge nerve gaps in rats. Immobilized gradients of LN-1 will be created in the agarose hydrogels by photochemical conjugation. Lipid-based microtubules (LMTs)-loaded with NGF will be embedded within these agarose scaffolds to slow-release NGF. A variable loading of LMTs into the agarose scaffold will be used to generate a diffuse gradient of NGF in the scaffolds. Thus, our anisotropic scaffolds will have gradients of both LN-1 and NGF, two important neurostimulatory molecules, to mimic *in vivo* conditions.

1.12. Animal Models for Nerve Regeneration

Traditionally, nerve regeneration studies have involved the use of various animal models, such as mice, rat, pigs, dogs, sheep and non-human primates. Rat or mice models are used initially to determine the efficacy of the various treatments. If the results are encouraging, they are followed by experiments with larger animal models. For PNS regeneration studies, the most commonly studied nerve models are the sciatic nerve and its branches, the tibial and the peroneal nerves. Other models include the cavernous nerve, femoral nerve and the facial nerve. The most common nerve injury model is the single-anastomosis model where the injury and repair are done on one sciatic nerve and the contralateral sciatic nerve is used as a control. This model is useful when the nerve gap is not more than 20-mm. The second version is the cross (double) anastomosis model, where both ipsilateral and contralateral sciatic nerves are transected; the proximal end of the right sciatic nerve is then sutured to one end of an implanted tube and the distal end of the left sciatic nerve is inserted into the other end of the tube [84]. This model allows the study of gaps in excess of 25-mm. Although very convenient, the rodent models suffer from the serious drawback that they present only short nerve gaps for regeneration studies. In order for a regeneration technique to be successfully applied in clinical trials, the nerve gap model has to be more than 40 mm in length. To create a long nerve gap model, rabbits [41], cats [114], dogs [88], sheep [75] and non-human primates [4] have been used. The large animal models are an important intermediary step before clinical application of experimental therapeutic approaches.

1.13. Conclusions

In spite of significant advances in research to the development of synthetic NGCs, nerve autografts are still considered the first-choice strategy for nerve repair, especially in the case of long nerve gaps. However, complete functional recovery is seldom observed with autografts. Using autografts generally results in a good recovery of sensory functions but negligible return of motor functions. Also, autografts have limited availability. Hence, there is lot of interest in developing synthetic nerve guidance channels as alternatives to autografts. The properties of guidance channels can be readily modified to enhance their function. Moreover, guidance channels can be incorporated with cells, neurotrophic factors, ECM proteins, and scaffolds, to enhance their performance. However, guidance channels still have not been able to match the performance of autografts. Ongoing rapid advances in cell biology, cell culture techniques, genetic engineering and biomaterials research, are likely to provide new tools to improve regeneration using NGCs, and the day an engineered construct performs as well as autografts may be near.

In order to improve upon the performance of nerve guidance channels, we propose the design of a novel agarose-based 3D scaffold with directional cues for regeneration *in vivo*. New techniques to create immobilized gradients of LN-1, and diffuse gradients of NGF have been proposed. These scaffolds will create regenerative conditions around the injured axons that are closer to physiological conditions. These scaffolds do not use cells, and hence are easy to fabricate and implant *in vivo*. If successful, these scaffolds may provide a significant advance in our efforts to promote nerve regeneration.

1.14. References

[1] Adams, D.N., Kao, E.Y., Hypolite, C.L., Distefano, M.D., Hu, W.S. and Letourneau, P.C., Growth cones turn and migrate up an immobilized gradient of the laminin IKVAV peptide, *J Neurobiol*, 62 (2005) 134-47.

[2] Aebischer, P., Guenard, V. and Brace, S., Peripheral nerve regeneration through blind-ended semipermeable guidance channels:effect of molecular weight cutoff, *J Neurosci*, 9 (1989a) 3590-3595.

[3] Aebischer, P., Salessiotis, A.N. and Winn, S.R., Basic fibroblast growth factor released from synthetic guidance channels facilitates peripheral nerve regeneration across long nerve gaps, *J Neurosci Res*, 23 (1989) 282-9.

[4] Ahmed, Z., Brown, R.A., Ljungberg, C., Wiberg, M. and Terenghi, G., Nerve growth factor enhances peripheral nerve regeneration in non-human primates, *Scand J Plast Reconstr Surg Hand Surg*, 33 (1999) 393-401.

[5] Albert, E., Einige Operationen an Nerven, *Wien Med*, 26 (1885) 1285.

[6] Amoh, Y., Li, L., Campillo, R., Kawahara, K., Katsuoka, K., Penman, S. and Hoffman, R.M., Implanted hair follicle stem cells form Schwann cells that support repair of severed peripheral nerves, *Proc Natl Acad Sci U S A*, 102 (2005) 17734-8.

[7] Ansselin, A.D., Fink, T. and Davey, D.F., Peripheral nerve regeneration through nerve guides seeded with adult Schwann cells, *Neuropathol Appl Neurobiol*, 23 (1997) 387-98.

[8] Barde, Y.A., Trophic factors and neuronal survival, *Neuron*, 2 (1989) 1525-34.

[9] Baron-van Evercooren, A., Kleinman, H.D., Ohno, S., Marangos, P., Schwartz, J.P. and Dubois-Dalcq, M.E., Nerv growth factor, laminin and fibronectin promote nerve growth in human fetal sensory ganglia cultures, *J Neurosci. Res.* (1982) 179-183.

[10] Battiston, B., Tos, P., Cushway, T.R. and Geuna, S., Nerve repair by means of vein filled with muscle grafts I. Clinical results, *Microsurgery*, 20 (2000) 32-36.

[11] Bellamkonda, R.V., Ranieri, J.P., Bouche, N. and Aebischer, P., Hydrogel-based three-dimensional matrix for neural cells, *J Biomed Mater Res*, 29 (1995) 663-71.

[12] Bonhoeffer, F. and Huf, J., In vitro experiments on axon guidance demonstrating an anterior-posterior gradient on the tectum, *Embo J*, 1 (1982) 427-31.

[13] Bonner, J. and O'Connor, T.P., The permissive cue laminin is essential for growth cone turning in vivo, *J Neurosci*, 21 (2001) 9782-91.

[14] Brandt, J., Dahlin, L.B. and Lundborg, G., Autologous tendons used as grafts for bridging peripheral nerve defects, *J Hand Surg [Br]*, 24 (1999) 284-90.

[15] Bryan, D.J., Wang, K.-K. and Chakalis-Haley, D.P., Effect of Schwann cells in the enhancement of peripheral nerve regeneration, *J. Reconstr. Microsurg.*, 12 (1996) 439-446.

[16] Bunge, R.P., Expanding roles for the Schwann cell: ensheathment, myelination, trophism and regeneration, *Curr Opin Neurobiol*, 3 (1993) 805-9.

[17] Cajal, S.R.y., Degeneration and regeneration of the nervous system, *Hafner* (1928).

[18] Cao, X. and Shoichet, M.S., Investigating the synergistic effect of combined neurotrophic factor concentration gradients to guide axonal growth, *Neuroscience*, 122 (2003) 381-9.

[19] Chen, Y.S., Hsieh, C.L., Tsai, C.C., Chen, T.H., Cheng, W.C., Hu, C.L. and Yao, C.H., Peripheral nerve regeneration using silicone rubber chambers filled with collagen, laminin and fibronectin, *Biomaterials*, 21 (2000) 1541-7.

[20] Chiu, D.T., Janecka, I., Krizek, T.J., Wolff, M. and Lovelace, R.E., Autogenous vein graft as a conduit for nerve regeneration, *Surgery*, 91 (1982) 226-33.

[21] Chiu, D.T., Lovelace, R.E., Yu, L.T., Wolff, M., Stengel, S., Middleton, L., Janecka, I.P., and Krizek, T.J., Comparative electrophysiological evaluation of nerve grafts and autogenous vein grafts as nerve conduits: an experimental study, *J. Reconstr. Microsurg.*, 4 (1988) 303-309.

[22] Chiu, D.T. and Strauch, B., A prospective clinical evaluation of autogenous vein grafts used as a nerve conduit for distal sensory nerve defects of 3 cm or less, *Plast Reconstr Surg*, 86 (1990) 928-34.

[23] Choi, B.H., Zhu, S.J., Kim, B.Y., Huh, J.Y., Lee, S.H. and Jung, J.H., Transplantation of cultured bone marrow stromal cells to improve peripheral nerve regeneration, *Int J Oral Maxillofac Surg*, 34 (2005) 537-42.

[24] Choi, B.H., Zhu, S.J., Kim, S.H., Kim, B.Y., Huh, J.H., Lee, S.H. and Jung, J.H., Nerve repair using a vein graft filled with collagen gel, *J Reconstr Microsurg*, 21 (2005) 267-72.

[25] Clark, P., Britland, S. and Connolly, P., Growth cone guidance and neuron morphology on micropatterned laminin surfaces, *J Cell Sci*, 105 (Pt 1) (1993) 203-12.

[26] Cornbrooks, C.J., Carey, D.J., McDonald, J.A., Timpl, R. and Bunge, R.P., In vivo and in vitro observations on laminin production by Schwann cells, *Proc Natl Acad Sci U S A*, 80 (1983) 3850-4.

[27] Dahlin, L.B. and Lundborg, G., Use of tubes in peripheral nerve repair, *Neurosurg Clin N Am*, 12 (2001) 341-52.

[28] Derby, A., Engleman, V.W., Frierdich, G.E., Neises, G., Rapp, S.R. and Roufa, D.G., Nerve growth factor facilitates regeneration across nerve gaps: morphological and behavioral studies in rat sciatic nerve, *Exp Neurol*, 119 (1993) 176-91.

[29] Dertinger, S.K., Jiang, X., Li, Z., Murthy, V.N. and Whitesides, G.M., Gradients of substrate-bound laminin orient axonal specification of neurons, *Proc Natl Acad Sci U S A*, 99 (2002) 12542-7.

[30] Dillon, G.P., Yu, X. and Bellamkonda, R.V., The polarity and magnitude of ambient charge influences three-dimensional neurite extension from DRGs, *J Biomed Mater Res*, 51 (2000) 510-9.

[31] Dubey, N., Letourneau, P.C. and Tranquillo, R.T., Neuronal contact guidance in magnetically aligned fibrin gels: effect of variation in gel mechano-structural properties, *Biomaterials*, 22 (2001) 1065-75.

[32] Edelman, E.R., Mathiowitz, E., Langer, R. and Klagsbrun, M., Controlled and modulated release of basic fibroblast growth factor, *Biomaterials*, 12 (1991) 619-26.

[33] Edgar, D., Nerve growth factors and molecules of the extracellular matrix in neuronal development, *J Cell Sci Suppl*, 3 (1985) 107-13.

[34] English, A.W., Meador, W. and Carrasco, D.I., Neurotrophin-4/5 is required for the early growth of regenerating axons in peripheral nerves., *Eur. J. Neurosci.*, 21 (2005) 2624-34.

[35] Esch, T., Lemmon, V. and Banker, G., Local presentation of substrate molecules directs axon specification by cultured hippocampal neurons, *J Neurosci*, 19 (1999) 6417-6426.

[36] Evans, G.R., Peripheral nerve injury: a review and approach to tissue engineered constructs, *Anat Rec*, 263 (2001) 396-404.

[37] Fawcett, J.W.a.K., R.J., Peripheral nerve regeneration, *Ann. Rev. Neurosci.*, 13 (1990) 43-60.

[38] Ferrari, F., De Castro Rodrigues, A., Malvezzi, C.K., Dal Pai Silava, M. and Padvoni, C.R., Inside-out vs. standard vein graft to repair a sensory nerve in rats, *Anat Rec*, 256 (1999) 227-232.

[39] Fu, S.Y. and Gordon, T., The cellular and molecular basis of peripheral nerve regeneration, *Mol Neurobiol*, 14 (1997) 67-116.

[40] Garcia-Alonso, L., Fetter, R.D. and Goodman, C.S., Genetic analysis of Laminin A in Drosophila: extracellular matrix containing laminin A is required for ocellar axon pathfinding, *Development*, 122 (1996) 2611-21.

[41] Geuna, S., Tos, P., Battiston, B. and Giacobini-Robecchi, M.G., Bridging peripheral nerve defects with muscle-vein combined guides, *Neurol Res*, 26 (2004) 139-44.

[42] Glasby, M.A., Gschmeissner, S.G., Hitchcock, R.J., and Huang, C.L, The dependence of nerve regeneration through muscle grafts in the rat on the availability an dorientation of basement membrane, *J. Neurocytol.*, 15 (1986) 497-510.

[43] Glazner, G.W., Lupien, S., Miller, J.A. and Ishii, D.N., Insulin-like growth factor II increases the rate of sciatic nerve regeneration in rats, *Neuroscience*, 54 (1993) 791-7.

[44] Gluck, T., Ueber Neuroplastik auf dem Wege der Transplantation, *Arch Klin Chir*, 25 (1880) 606-616.

[45] Gospodarowicz, D., Ferrara, N., Schweigerer, L. and Neufeld, G., Structural characterization and biological functions of fibroblast growth factor, *Endocr Rev*, 8 (1987) 95-114.

[46] Griffin, C.G. and Letourneau, P.C., Rapid retraction of neurites by sensory neurons in response to increased concentrations of nerve growth factor, *J Cell Biol*, 86 (1980) 156-61.

[47] Groves, M.L., McKeon, R., Werner, E., Nagarsheth, M., Meador, W. and English, A.W., Axon regeneration in peripheral nerves is enhanced by proteoglycan degradation, *Exp Neurol*, 195 (2005) 278-292.

[48] Guenard, V., Kleitman, N., Morrissey, T.K., Bunge, R.P. and Aebischer, P., Syngeneic Schwann cells derived from adult nerves seeded in semipermeable guidance channels enhance peripheral nerve regeneration, *J Neurosci*, 12 (1992) 3310-20.

[49] Gundersen, R.W. and Barrett, J.N., Neuronal chemotaxis: chick dorsal-root axons turn toward high concentrations of nerve growth factor, *Science*, 206 (1979) 1079-80.

[50] Gundersen, R.W. and Barrett, J.N., Characterization of the turning response of dorsal root neurites toward nerve growth factor, *J Cell Biol*, 87 (1980) 546-54.

[51] Hadlock, T., Sundback, C., Hunter, D., Cheney, M. and Vacanti, J.P., A polymer foam conduit seeded with Schwann cells promotes guided peripheral nerve regeneration, *Tissue Eng*, 6 (2000) 119-27.

[52] Halfter, W., The behavior of optic axons on substrate gradients of retinal basal lamina proteins and merosin, *J Neurosci*, 16 (1996) 4389-401.

[53] Hall, S.M., The effect of inhibiting Schwann cell mitosis on the re-innervation of acellular autografts in the peripheral nervous system of the mouse, *Neuropathol Appl Neurobiol*, 12 (1986) 401-14.

[54] Heumann, R., Korsching, S., Bandtlow, C. and Thoenen, H., Changes of nerve growth factor synthesis in nonneuronal cells in response to sciatic nerve transection, *J Cell Biol*, 104 (1987) 1623-31.

[55] Hoffman, D., Wahlberg, L. and Aebischer, P., NGF released from a polymer matrix prevents loss of ChAT expression in basal forebrain neurons following a fimbria-fornix lesion, *Exp Neurol*, 110 (1990) 39-44.

[56] http://www.axogeninc.com, AxoGen Inc., October 10, 2006.

[57] Hudson, T.W., Evans, G.R. and Schmidt, C.E., Engineering strategies for peripheral nerve repair, *Orthop Clin North Am*, 31 (2000) 485-98.

[58] Hynes, R.O., Integrins: a family of cell surface receptors, *Cell*, 48 (1987) 549-54.

[59] Ide, C., Peripheral nerve regeneration, *Neurosci Res*, 25 (1996) 101-21.

[60] Jansen, K., van der Werff, J.F., van Wachem, P.B., Nicolai, J.P., de Leij, L.F. and van Luyn, M.J., A hyaluronan-based nerve guide: in vitro cytotoxicity, subcutaneous tissue reactions, and degradation in the rat, *Biomaterials*, 25 (2004) 483-9.

[61] Jing, S., Tapley, P. and Barbacid, M., Nerve growth factor mediates signal transduction through trk homodimer receptors, *Neuron*, 9 (1992) 1067-79.

[62] Kam, L., Shain, W., Turner, J.N. and Bizios, R., Axonal outgrowth of hippocampal neurons on micro-scale networks of polylysine-conjugated laminin, *Biomaterials*, 22 (2001) 1049-54.

[63] Kapur, T.A. and Shoichet, M.S., Immobilized concentration gradients of nerve growth factor guide neurite outgrowth, *J Biomed Mater Res A*, 68 (2004) 235-43.

[64] Keilhoff, G., Pratsch, F., Wolf, G. and Fansa, H., Bridging extra large defects of peripheral nerves: possibilities and limitations of alternative biological grafts from acellular muscle and Schwann cells, *Tissue Eng*, 11 (2005) 1004-14.

[65] Kerkhoff, H. and Jennekens, F.G., Peripheral nerve lesions: the neuropharmacological outlook, *Clin Neurol Neurosurg*, 95 Suppl (1993) S103-8.

[66] Keynes, R. and Cook, G.M., Axon guidance molecules, *Cell*, 83 (1995) 161-9.

[67] Kirk, E.G. and D., L., Fascial tubulization in the repair of nerve defects, *JAMA*, 65 (1915) 486-492.

[68] Kiyotani, T., Nakamura, T., Shimuzu, Y., Endo, K., Experimental study of nerve regeneration in a biodegradable tube made from collagen and polyglycolic acid, *ASAIO J.*, 41 (1995) M657-661.

[69] Kleinman, H.K., Ogle, R.C., Cannon, F.B., Little, C.D., Sweeney, T.M. and Luckenbill-Edds, L., Laminin receptors for neurite formation, *Proc Natl Acad Sci U S A*, 85 (1988) 1282-6.

[70] Kline, D.G., Kim, D., Midha, R., Harsh, C. and Tiel, R., Management and results of sciatic nerve injuries: a 24-year experience, *J Neurosurgery*, 89 (1998) 13-23.

[71] Krewson, C.E., Klarman, M.L. and Saltzman, W.M., Distribution of nerve growth factor following direct delivery to brain interstitium, *Brain Res*, 680 (1995) 196-206.

[72] Kuffler, D.P., Isolated satellite cells of a peripheral nerve direct the growth of regenerating frog axons, *J Comp Neurol*, 249 (1986) 57-64.

[73] Labrador, R.O., Buti, M. and Navarro, X., Peripheral nerve repair: role of agarose matrix density on functional recovery, *Neuroreport*, 6 (1995) 2022-6.

[74] Labrador, R.O., Buti, M. and Navarro, X., Influence of collagen and laminin gels concentration on nerve regeneration after resection and tube repair, *Exp Neurol*, 149 (1998) 243-52.

[75] Lawson, G.M. and Glasby, M.A., Peripheral nerve reconstruction using freeze-thawed muscle grafts: a comparison with group fascicular nerve grafts in a large animal model, *J R Coll Surg Edinb.*, 43 (1998) 295-302.

[76] Lee, A.C., Yu, V.M., Lowe, J.B., 3rd, Brenner, M.J., Hunter, D.A., Mackinnon, S.E. and Sakiyama-Elbert, S.E., Controlled release of nerve growth factor enhances sciatic nerve regeneration, *Exp Neurol*, 184 (2003) 295-303.

[77] Letourneau, P.C., Condic, M.L. and Snow, D.M., Interactions of developing neurons with the extracellular matrix, *J Neurosci*, 14 (1994) 915-28.

[78] Levi-Montalcini, R., The nerve growth factor 35 years later, *Science*, 237 (1987) 1154-62.

[79] Levi, A.D., Sonntag, V.K., Dickman, C., Mather, J., Li, R.H., Cordoba, S.C., Bichard, B. and Berens, M., The role of cultured Schwann cell grafts in the repair of gaps within the peripheral nervous system of primates, *Exp. Neurol.*, 143 (1997) 25-36.

[80] Li, Y., Decherchi, P. and Raisman, G., Transplantation of olfactory ensheathing cells into spinal cord lesions restores breathing and climbing, *J Neurosci*, 23 (2003) 727-31.

[81] Liuzzi, F.J. and Tedeschi, B., Peripheral nerve regeneration, *Neurosurg Clin N Am*, 2 (1991) 31-42.

[82] Luckenbill-Edds, L., Laminin and the mechanism of neuronal outgrowth, *Brain Research Reviews*, 23 (1997) 1-27.

[83] Lundborg, G., *Nerve Injury and Repair*, Longman Group UK, New York, 1988.

[84] Lundborg, G., Dahlin, L.B., Danielsen, N., Gelberman, R.H., Longo, F.M., Powell, H.C. and Varon, S., Nerve regeneration in silicone chambers: influence of gap length and of distal stump components, *Exp Neurol*, 76 (1982) 361-75.

[85] Mackinnon, S.E., Dellon, A.L., Clinical nerve reconstruction with a bioabsorbable polyglycolic acid tube, *Plast. Reconstr. Surg.*, 85 (1990) 419-424.

[86] Madison, R.D., Da Silva, C.F. and Dikkes, P., Entubulation repair with protein additives increases the maximum nerve gap distance successfully bridged with tubular prostheses, *Brain Res*, 447 (1988) 325-34.

[87] Martini, R., Expression and functional roles of neural cell surface molecules and extracellular matrix components during development and regeneration of peripheral nerves, *J Neurocytol*, 23 (1994) 1-28.

[88] Matsumoto, K., Ohnishi, K., Kiyotani, T., Sekine, T., Ueda, H., Nakamura, T., Endo, K. and Shimizu, Y., Peripheral nerve regeneration across an 80-mm gap bridged by a polyglycolic acid (PGA)-collagen tube filled with laminin-coated collagen fibers: a histological and electrophysiological evaluation of regenerated nerves, *Brain Res*, 868 (2000) 315-28.

[89] Matsuzawa, M., Tokumitsu, S., Knoll, W. and Leisi, P., Molecular gradeint along the axon pathway is not required for directional axon growth, *J Neurosci Res*, 53 (1998) 114-124.

[90] McCormack, M.L., Goddard, M., Guenard, V. and Aebischer, P., Comparison of dorsal and ventral spinal root regeneration through semipermeable guidance channels, *J Comp Neurol*, 313 (1991) 449-56.

[91] McKenna, M.P. and Raper, J.A., Growth cone behavior on gradients of substratum bound laminin, *Dev Biol*, 130 (1988) 232-236.

[92] Meek, M.F., Varejo, A.S., Geuna, S., Muscle grafts and alternatives for nerve repair, *J. Oral Maxillofac. Surg.*, 60 (2002) 1095-96.

[93] Meyer, R.S., Abrams, R.A., Botte, M.J., Davey, J.P. and Bodine-Fowler, S.C., Functional recovery following neurorrhaphy of the rat sciatic nerve by epineurial repair compared with tubulization, *J Orthop Res*, 15 (1997) 664-9.

[94] Millesi, H., Meissl, G. and Berger, A., The interfascicular nerve-grafting of the median and ulnar nerves, *J Bone Joint Surg Am.*, 54 (1972) 7727-7750.

[95] Nakahara, Y., Gage, F.H. and Tuszynski, M.H., Grafts of fibroblasts genetically modified to secrete NGF, BDNF, NT-3, or basic FGF elicit differential responses in the adult spinal cord, *Cell Transplant*, 5 (1996) 191-204.

[96] Pagnotta, A., Tos, P., Fornaro, M., Gigante, A., Geuna, S., Neurtrophins and their receptors in early axonal regeneration along muscle-vein-combined grafts, *Microsurgery*, 22 (2002) 300-3.

[97] Payr, E., Beitrage zur Technik der Blutgefass und Nervennaht nebst Mittheilungen uber die Vervendung eines resorbibaren Metalles in der Chirurgie, *Arch Klin Chir*, 62 (1900) 67.

[98] Raivich, G. and Kreutzberg, G.W., Peripheral nerve regeneration: role of growth factors and their receptors, *Int J Dev Neurosci*, 11 (1993) 311-24.

[99] Reichardt, L.F. and Tomaselli, K.J., Extracellular matrix molecules and their receptors: functions in neural development, *Annu Rev Neurosci*, 14 (1991) 531-70.

[100] Rich, K.M., Luszczynski, J.R., Osborne, P.A. and Johnson, E.M., Jr., Nerve growth factor protects adult sensory neurons from cell death and atrophy caused by nerve injury, *J Neurocytol*, 16 (1987) 261-8.

[101] Rivas, R.J., Burmeister, D.W. and Goldberg, D.J., Rapid effects of laminin on the growth cone, *Neuron*, 8 (1992) 107-115.

[102] Rogers, S.L., Letourneau, P.C., Palm, S.L., McCarthy, J. and Furcht, L.T., Neurite extension by central and peripheral nervous system neurons in response to substratun-bound fibronectin and laminin, *Dev Biol*, 98 (1983) 212-220.

[103] Rudolph, A.S., Stilwell, G., Cliff, R.O., Kahn, B., Spargo, B.J., Rollwagen, F. and Monroy, R.L., Biocompatibility of lipid microcylinders: effect on cell growth and antigen presentation in culture, *Biomaterials*, 13 (1992) 1085-92.

[104] Sakiyama-Elbert, S.E. and Hubbell, J.A., Development of fibrin derivatives for controlled release of heparin-binding growth factors, *J Control Release*, 65 (2000) 389-402.

[105] Saneinejad, S. and Shoichet, M.S., Patterned glass surfaces direct cell adhesion and process outgrowth of primary neurons of the central nervous system, *J Biomed Mater Res*, 42 (1998) 13-9.

[106] Satou, T., Nishida, S., Hiruma, S., Tanji, K., Takahashi, M., Fujita, S., Mizuhara, Y., Akai, F. and Hashimoto, S., A morphological study on the effects of collagen gel matrix on regeneration of severed rat sciatic nerve in silicone tubes, *Acta Pathol Jpn*, 36 (1986) 199-208.

[107] Schanen-King, C., Nel, A., Williams, L.K. and Landreth, G., Nerve growth factor stimulates the tyrosine phosphorylation of MAP2 kinase in PC12 cells, *Neuron*, 6 (1991) 915-22.

[108] Seckel, B.R., Jones, D., Hekimian, K.J., Wang, K.K., Chakalis, D.P. and Costas, P.D., Hyaluronic acid through a new injectable nerve guide delivery system enhances peripheral nerve regeneration in the rat, *J Neurosci Res*, 40 (1995) 318-24.

[109] Sendtner, M., Holtmann, B., Kolbeck, R., Thoenen, H. and Barde, Y.A., Brain-derived neurotrophic factor prevents the death of motoneurons in newborn rats after nerve section, *Nature*, 360 (1992) 757-9.

[110] Song, H. and Poo, M., The cell biology of neuronal navigation, *Nat Cell Biol*, 3 (2001) E81-8.

[111] Spargo, B.J., Cliff, R.O., Rollwagen, F.M. and Rudolph, A.S., Controlled release of transforming growth factor-beta from lipid-based microcylinders, *J Microencapsul*, 12 (1995) 247-54.

[112] Strauch, B., Rodriguez, D.M., Diaz, J., Yu, H.L., Kaplan, G. and Weinstein, D.E., Autologous Schwann cells drive regeneration through a 6-cm autogenous venous nerve conduit, *J Reconstr Microsurg*, 17 (2001) 589-595.

[113] Suematsu, N., Tubulation of the peripheral nerve gap: Its history and possibility, *Microsurgery*, 10 (1989) 1-74.

[114] Suzuki, Y., Tanihara, M., Ohnishi, K., Suzuki, K., Endo, K. and Nishimura, Y., Cat peripheral nerve regeneration across 50 mm gap repaired with a novel nerve guide composed of freeze-dried alginate gel, *Neurosci Lett*, 259 (1999) 75-8.

[115] Terzis, J., Faibisoff, B. and Williams, B., The nerve gap: suture under tension vs. graft, *Plast Reconstr Surg*, 56 (1975) 166-70.

[116] Tessier-Lavigne, M. and Goodman, C.S., The molecular biology of axon guidance, *Science*, 274 (1996) 1123-33.

[117] Thanos, P.K., Okajima, S. and Terzis, J.K., Utrastructure and cellualr biology of nerve regeneration, *J. Reconstr. Microsurg.*, 14 (1998) 423-36.

[118] Thoenen, H., Barde, Y.A., Davies, A.M. and Johnson, J.E., Neurotrophic factors and neuronal death, *Ciba Found Symp*, 126 (1987) 82-95.

[119] Timple, R., Immunological studies on collagen. In G.N. Ramachandran and A.H. Reddi (Eds.), *Biochemistry of collagen*, Plenum Press, New York, 1976, pp. 319-375.

[120] Tresco, P., Tissue engineering strategies for nervous system repair, *Progress in Brain Research*, 128 (2000) 349-363.

[121] Uzman, B.G., Villegas, G.M., Mouse sciatic nerve regeneration through semi-permeable tubes: a quantitative model, *J Neurosci*, 9 (1983) 325-338.

[122] Valentini, R.F. and Aebischer, P., Strategies for the engineering of peripheral nervous tissue regeneration. In L.R. Lanza RP, Chick WL (Ed.), *Principles of Tissue Engineering*, R.G.Landes Company, Austin, 1997, pp. 671-684.

[123] Valentini, R.F., Aebischer, P., Winn, S.R. and Galletti, P.M., Collagen- and laminin-containing gels impede peripheral nerve regeneration through semipermeable nerve guidance channels, *Exp Neurol*, 98 (1987) 350-6.

[124] Walton, R.L., Brown, R.E., Matory, W.E. Jr, Borah, G.L., Dolph, J.L., Autogenous vein graft repair of digital nerve defects in the finger: a retrospective clinical study, *Plast. Reconstr. Surg.*, 84 (1989) 944-49.

[125] Wang, G.Y., Hirai, K. and Shimada, H., The role of laminin, a component of Schwann cell basal lamina, in rat sciatic nerve regeneration within antiserum-treated nerve grafts, *Brain Res*, 570 (1992a) 116-25.

[126] Wang, G.Y., Hirai, K., Shimada, H., Taji, S. and Zhong, S.Z., Behavior of axons, Schwann cells and perineurial cells in nerve regeneration within transplanted nerve grafts: effects of anti-laminin and anti-fibronectin antisera, *Brain Res*, 583 (1992b) 216-26.

[127] Wang, K.K., Costas, P.D., Bryan, D.J., Jones, D.S. and Seckel, B.R., Inside-out vein graft promotes improved nerve regeneration in rats, *J Reconstr Microsurg*, 14 (1993) 608-618.

[128] Wang, K.K., Nemeth, I.R., Seckel, B.R., Chakalis-Haley, D.P., Swann, D.A., Kuo, J.W., Bryan, D.J. and Cetrulo, C.L., Jr., Hyaluronic acid enhances peripheral nerve regeneration in vivo, *Microsurgery*, 18 (1998) 270-5.

[129] Weiss, P. and Taylor, A.C., Guides for nerve regeneration across nerve gaps, *J Neurosurg*, 3 (1946) 275-82.

[130] Williams, L.R., Longo, F.M., Powell, H.C., Lundborg, G. and Varon, S., Spatial-temporal progress of peripheral nerve regeneration within a silicone chamber: parameters for a bioassay, *J Comp Neurol*, 218 (1983) 460-70.

[131] Xu, X.M., Zhang, S.X., Li, H., Aebischer, P., and Bunge, M.B., Regrowth of axons into distal spinal cord through a Schwann cell seeded mini-channel implanted into hemisected adult rat spinal cord, *Eur. J. Neurosci.*, 11 (1999) 1723-1740.

[132] Yu, X. and Bellamkonda, R.V., Tissue-engineered scaffolds are effective alternatives to autografts for bridging peripheral nerve gaps, *Tissue Eng*, 9 (2003) 421-30.

[133] Yu, X., Dillon, G.P. and Bellamkonda, R.B., A laminin and nerve growth factor-laden three-dimensional scaffold for enhanced neurite extension, *Tissue Eng*, 5 (1999) 291-304.

[134] Zhang, F., Blain, B., Beck, J., Zhang, J., Chen, Z., Chen, Z.W. and Leineaweaver, W.C., Autogenous venous graft with one-stage prepared Schwann cells as a conduit for repair of long segmental nerve defects, *J Reconstr Microsurg*, 18 (2002) 295-300.

[135] Zhao, Q., Dahlin, L.B., Kanje, M. and Lundborg, G., Specificity of muscle reinnervation following repair of the transected sciatic nerve. A comparative study of different repair techniques in the rat, *J Hand Surg [Br]*, 17 (1992) 257-61.

[136] Zuo, J., Neubauer, D., Graham, J., Krekoski, C.A., Ferguson, T.A. and Muir, D., Regeneration of axons after nerve transection repair is enhanced by degradation of chondroitin sulfate proteoglycan, *Exp Neurol*, 176 (2002) 221-8.

Chapter 2: **ANISOTROPIC SCAFFOLDS FACILITATE ENHANCED NEURITE EXTENSION *IN VITRO***

This work has been published in the Journal of Biomedical Materials Research (A), Aug. 2006; 78(2):213-21.

Tissue engineering (TE) techniques to enhance nerve regeneration following nerve damage have had limited success in matching the performance of autografts across short nerve gaps (< 10mm). For regeneration over longer nerve gaps, TE techniques have been less successful than autografts. Most engineered scaffolds do not present directional cues to the regenerating nerves. In our efforts to design a TE scaffold to replace the autograft, we hypothesize that anisotropic hydrogel scaffolds with gradients of a growth promoting glycoprotein, laminin-1 (LN-1), may promote directional neurite extension and enhance regeneration. In this study we report the engineering of three-dimensional (3D) agarose scaffolds with photoimmobilized gradients of LN-1 of differing slopes. Dorsal root ganglia (DRG) from chicken embryos were cultured in the agarose scaffolds and their neurite extension rate was determined. DRG neurite extension rates were significantly higher in the anisotropic scaffolds, with a maximal growth rate in an anisotropic scaffold twice that of the maximal growth rate in isotropic scaffolds of LN-1. We suggest that these anisotropic scaffolds, presenting an optimal gradient of LN-1, may significantly impact nerve regeneration. Such anisotropic scaffolds may represent a new generation of tissue-engineered materials for enhanced tissue or nerve regeneration.

2.1. Introduction

Injuries to the nervous system can lead to loss of sensory and motor functions and result in debilitating diseases. In cases where the peripheral nerve is damaged, techniques to repair damaged peripheral nerves include coaptation by suturing the two ends of a severed peripheral nerve together. When the gaps are large, autografts are used as the bridging materials. However, gaps greater than 3 mm cannot be repaired through coaptation and procuring autografts is often difficult due to lack of availability. This has necessitated the pursuit of alternatives to the use of autografts [132,151]. One such alternative approach is to design a 3D scaffold containing growth-promoting extracellular matrix (ECM) proteins, such as laminin-1 (LN-1), to promote peripheral nerve regeneration [107,163]. Laminins (LN) are a family of ECM-bound glycoproteins found in the basement membrane of most tissues, and are expressed in the developing nervous system as well as the adult peripheral nervous system (PNS) [120]. Several studies have shown LN as potent promoters of neurite growth in several types of cultured neurons [13,124,126], as directional cues for axonal growth *in vitro* [1,32,39,66], and as important guidance molecules for steering and path-finding for developing axons *in vivo* [17,53].

The ability of LN to stimulate neurite outgrowth *in vitro,* along with the known role of LN in the developing nervous system [79,135], has led to the hypothesis that LN will contribute positively to peripheral nerve regeneration after injury. Even though regeneration in the PNS is more extensive than regeneration in the central nervous system (CNS), successful regeneration over large nerve gaps in the PNS is rare.

During development, the guidance of growth cones to their target tissues is mediated by multiple factors, such as chemoattractant, chemorepellant, contact-attractive and contact-repulsive molecules [143], and some of these growth cone guidance cues may be distributed in a graded fashion [16]. There is considerable evidence for gradients of soluble factors orienting the locomotion of non-neuronal cells and growth cones.[25,93] Although the influence of gradients of soluble substances on neuronal behavior has been extensively studied and has been used to unravel the molecular and cellular mechanisms of axonal guidance [135], much less is known about gradients of substrate bound substances [11], such as LN. We hypothesize that anisotropic scaffolds presenting a gradient of LN may enhance neurite extension relative to isotropic scaffolds, potentially leading to higher success rates in large nerve gap regeneration.

Since, appropriate substrate-bound and diffusible agents can influence nerve growth, biomaterials can be used to manipulate physical, chemical and biological cues around the injured neurons to control cellular responses and facilitate regeneration. Agarose is a polysaccharide derived from red agar, and is widely used in gel electrophoresis and gel chromatography. SeaPrep® agarose solutions gel at a low temperature (17°C) and are capable of thermo-reversibly changing back to liquid state when heated above 50°C. Therefore once gelled, it remains as a gel at physiological temperatures. SeaPrep® agarose gel is transparent allowing cells cultured in it to be easily viewed under the microscope. SeaPrep® agarose has many functional groups, which can be modified to attach proteins and other molecules to it. SeaPrep® agarose hydrogels have been shown to support neurite extension from a variety of neurons in a non-immunogenic manner [15,40,41,164]. Agarose gels also allow various molecules to be covalently linked to the gels through various functional groups on the polysaccharide chains. For example, LN-1 peptides or the full protein can be covalently coupled to

SeaPrep® agarose gels to enhance their ability to support neurite extension [164]. Although agarose gels support neurite growth on their own, coupling of LN-1 significantly enhances the ability of the gels to promote neurite extension. This makes agarose hydrogel a suitable 3D matrix for culturing neurons.

In this study we report the fabrication of anisotropic 3D agarose hydrogel scaffolds with gradients of coupled LN-1 and examine the effects of these gradients of LN-1 on neurite extension in 3D. We suggest that these scaffolds represent a new generation of tissue engineered scaffolds with built in directional cues for enhanced nerve regeneration.

2.2. Materials and Methods

2.2.1. Generation of isotropic and anisotropic 3D hydrogel scaffolds

2.2.1.1. Isotropic LN-1 scaffolds generated using photochemistry

Photochemical coupling was used to immobilize soluble, photosensitive LN-1 conjugate [23]. LN-1 (BD Biosciences, Bedford, MA) was coupled to the agarose scaffold using a photocrosslinker, Sulfo-SANPAH (Sulfosuccinimidyl-6-[4'-azido-2'-nitrophenylamino] hexanoate) (PIERCE, Rockford, IL). Sulfo-SANPAH is a hetro-bifunctional crosslinker with a Sulfo-NHS group, that chemically reacts with amine groups, and a photoreactive perfluoro arylazide group that replaces the hydrogen atom in C-H bonds when triggered with UV light, [149] as schematically depicted in Figure 2-1. When added to an aqueous solution containing LN-1, Sulfo-SANPAH immediately hydrolyzes and its reactive groups are exposed. The Sulfo-NHS group reacts with the amine group of LN-1 to form a LN-1-SANPAH conjugate. When the LN-1-SANPAH

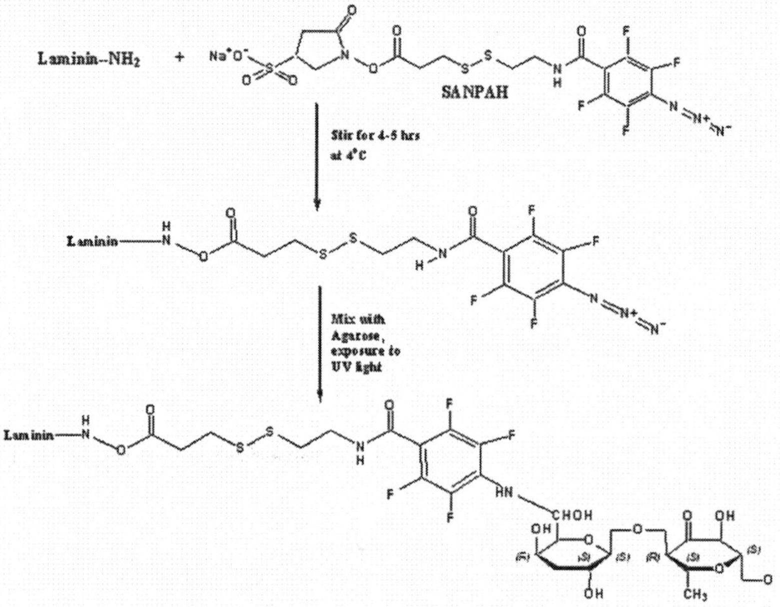

Figure 2-1: Schematic of photocrosslinking chemistry. LN-1 first reacts with Sulfo-SANPAH via its amine group. The LN-1-SANPAH conjugate is exposed to UV light, activating SANPAH and initiating reaction with C-H bonds in the agarose backbone resulting in the crosslinking of LN-1 to agarose.

conjugate is added to an agarose solution and exposed to UV light, the perfluoro arylazide group of SANPAH facilitates the coupling of LN-1 to agarose, which has an abundance of C-H bonds. Briefly, SANPAH was weighed and added to a solution of LN-1 in phosphate buffered saline (PBS; pH 7.4) solution so that SANPAH: LN-1 molar ratio is 100:1. The mixture was incubated with stirring at 500 RPM for 4-5 hrs at 4°C to allow conjugation of SANPAH with LN-1. The LN-1-SANPAH conjugate was dialyzed against PBS using Spectrum® dialysis blocks (MWCO 10 kDa, Spectrum Lab Inc., Rancho Dominguez, CA) to remove unreacted SANPAH molecules. To this conjugate solution, an equal volume of 2% (w/v) SeaPrep® agarose (BMA, Rockland, ME) solution was added. Agarose solution was prepared by dissolving agarose in PBS solution (pH 7.4) by heating at 60°C and stirring until the solution became clear. The resulting 1% (w/v) agarose scaffold with LN-1-SANPAH conjugate was then exposed to UV light from a BP-100AP lamp (100W, 365nm, 12.5 mW/cm^2 at a distance of 3", UVP, Upland, CA), held 3" from the sample, for 60 seconds, to trigger the coupling of LN-1 to the agarose solution via the SANPAH photocrosslinker. The solution was then allowed to gel by cooling it at 4°C for 20 minutes. Unbound LN-1 was removed from the gel by washing with PBS solution, with repeated changes, for 2 days. The gel was then heated, below 40°C, de-gelled and the amount of LN-1 coupled to the gel was determined by Bradford protein assay (BIO-RAD, Hercules, CA). As a control, agarose gel mixed with LN-1 (no SANPAH) was used under conditions identical to those for coupling SANPAH-LN-1.

2.2.1.2. Determination of efficiency of coupling LN-1 to agarose scaffold

In order to test the efficiency of photochemical coupling of LN-1 to agarose, different concentrations of LN-1-SANPAH conjugate were mixed with agarose solutions so as to obtain 1% (w/v) agarose solutions with a range of LN-1 concentrations - 85 μg/ml, 113.33 μg/ml and 170 μg/ml. These solutions were then exposed to UV light, at 3" from the UV lamp, for 60 seconds. The solutions were then allowed to gel by cooling at 4°C for 20 minutes. As a control, agarose scaffold mixed with LN-1 (no SANPAH) was used under conditions identical to those for coupling SANPAH-LN-1. The scaffolds were then washed with PBS solution for 2 days to remove any unbound LN-1. The amount of LN-1 coupled to these scaffolds was determined by Bradford protein assay.

2.2.1.3. Generation of anisotropic LN-1 concentrations in 3D agarose hydrogels

Eight-well chamber slides (Fisher Scientific, Pittsburg, PA) were used for making anisotropic LN-1 scaffolds and for culturing chick DRG ganglia. A 2.5 mm thick block of agarose hydrogel was formed in each well of the chamber slide by adding 1% (w/v) agarose solution between two custom-built teflon bars separated by a distance of 2.5 mm (Figure 2-2A). The chamber slide was placed at 4°C for 20 minutes to allow the agarose solution to gel. After the agarose gelled, the teflon bars were removed and the well was divided into two compartments by the agarose scaffold.

To visualize LN-1 gradient in agarose scaffold, LN-1 molecule was labeled with a fluorescent molecule, rhodamine, using EZ-Label Rhodamine protein labeling kit (PIERCE, Rockford, IL). One of the compartments was filled with 100 μl of LN-1-rhodamine solution (henceforth referred to as the "high" concentration compartment,

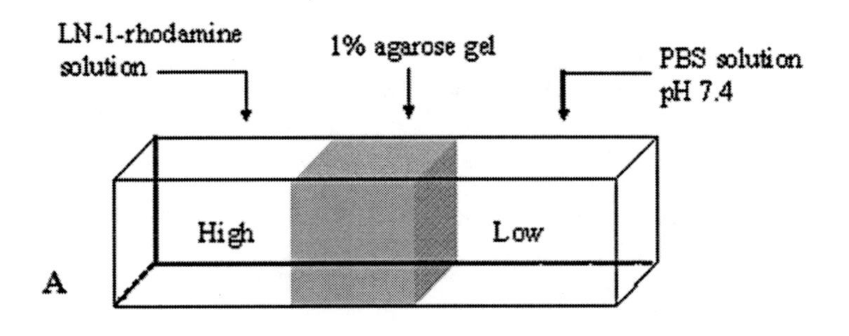

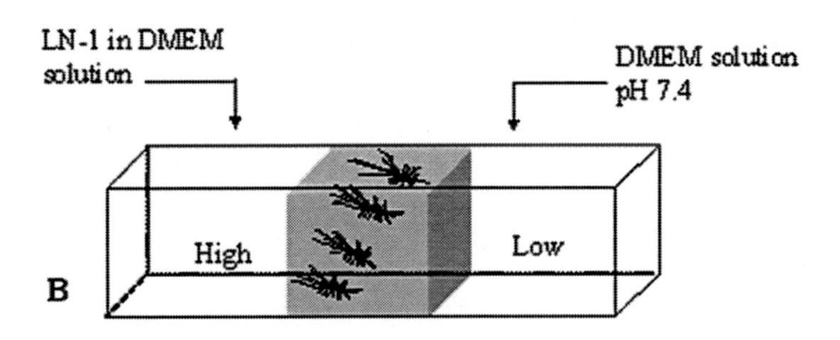

Figure 2-2: Schematic of experimental setup for gradient studies and DRG cultures. (A) To synthesize and characterize anisotropic LN-1 scaffolds an agarose scaffold block was prepared in the middle of a slide chamber well. LN-1 solution was added in the compartment marked "high" and PBS solution was added in the compartment marked "low". (B) For studying neurite extensions in LN-1 gradients, DRG were first mixed with agarose and then the LN-1 gradient was made.

Figure 2-2A) and the other compartment was filled with 100 μl of PBS solution with no LN-1 (henceforth referred to as the "low" concentration compartment, Figure 2-2A). The scaffolds were then stored in an incubator at 37°C and 95% humidity for 6 hours and LN-1 was allowed to diffuse through the agarose scaffold from the "high" concentration compartment to the "low" concentration compartment. After 6 hours, the solutions from the two compartments were removed and the scaffold block was immediately exposed to UV light from a BP-100AP lamp (12.5 mW/cm^2 at a distance of 3 inches) for 60 seconds to photoimmobilize LN-1-rhodamine. Uncoupled rhodamine-LN-1 was removed by washing the agarose scaffold block with PBS solution for 2 days. To test whether LN-1 was coupled to the scaffold in a gradient fashion, fluorescent images were captured using MicroFire™ digital camera (Olympus, Melville, NY) attached to an upright microscope (Model Axioskop 2, Carl Zeiss Inc., Germany). Images were taken at 3 different depths within the scaffolds, to check for uniformity of coupling through the thickness of the gel. The fluorescence intensity of the images was measured using ImagePro™ software (OPTRONICS, Coleta, CA). A relationship between fluorescence intensity and rhodamine-LN-1 concentrations was obtained by mixing known concentrations of rhodamine-LN-1 in 1% agarose scaffold and then measuring its fluorescence intensity.

2.2.2. Determining response of chick dorsal root ganglia neurites to LN-1 gradients

DRG culture in agarose hydrogel

Dorsal root ganglia (DRG) were harvested from 9-day old chicken embryos (E9) into PBS solution. Each DRG explant was split into 2-3 pieces and suspended in 1% agarose solution. An agarose scaffold block containing DRG was made as described above for plain agarose (Figure 2-2B). LN-1 solution diluted in Dulbecco's Modified Eagle Medium (DMEM) was added to the "high" concentration compartment and DMEM by itself was added to the "low" compartment. LN-1 was allowed to diffuse through the agarose scaffold for 6 hours to form a LN-1 gradient. The hydrogel was then exposed to UV light for 60 seconds at a distance of 3 inches. DMEM with 10% fetal bovine serum (FBS), 1% penicillin-streptomycin (PS) and 50 ng/ml nerve growth factor (NGF) was added to each of the wells. The cultures were then maintained at standard cell culture conditions of 37°C, 5% CO_2 and 95% humidity for 4-5 days. Neurite extension from DRG was visualized using an inverted microscope (Model Axiovert 200M, Carl Zeiss Inc., Germany). The lengths of specific, marked neurites were quantified on day 2, day 3 and day 4 of culture to help determine the rate of neurite extension. Extreme care was taken to measure the length of the same neurites for these measurements, using appropriate landmarks in the culture dish. The choice of neurites was not random, however, the longest neurites were identified by measuring their lengths at day 2, and subsequently these neurites were followed until day 4 of culture.

To test the influence on neurite outgrowth due to gradients of LN-1 as compared to uniform concentrations of LN-1, three conditions were studied. DRGs were cultured in 3D scaffolds with either uniform concentrations of LN-1 (0.2, 0.5, 0.8, 1.02, 3.4 and 10.21 μg/ml) or gradients of LN-1 (0.017 μg/ml/mm, 0.051 μg/ml/mm and 0.121 μg/ml/mm) or plain agarose (0 μg/ml). Neurites were considered as growing "up the gradient" when the angle (α) between the line connecting the point where it emerged from the DRG and the tip of the neurite, and the line following the steepest increase in the concentration of LN-1 (Δlaminin = maximum in Figure 2-3) was such that $-55^{\circ} < \alpha <$

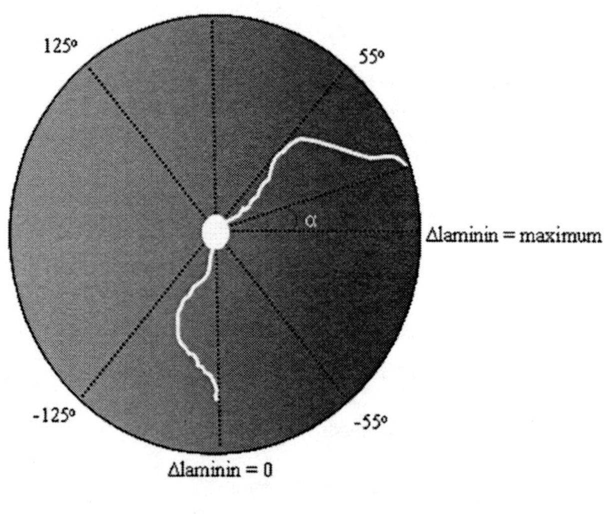

Figure 2-3: Schematic of an axonal process on a LN-1 gradient and how the angle α is defined. The neurites are considered as growing "up the gradient" if -55° < α < -55°, "down the gradient" if -125° < α < 125°, and "perpendicular to the gradient" if 55° < α < 125° and -125° < α < 125°.

55°. The 55° angle was selected because the neurites that grew within $\pm 55^o$ experienced 50% of the maximum slope of change in LN-1 concentration (Δlaminin) of that gradient. The neurites were considered as growing "perpendicular to the gradient" if $55^o < \alpha < 125^o$ and $-125^o < \alpha < -55^o$; and neurites were considered as growing "down the gradient" if they were extending in the direction of decreasing concentration of LN-1 ($125^o < \alpha < -125^o$). A schematic illustration of these definitions is shown in Figure 2-3. One-way ANOVA test was used for statistical analysis and a p-value of less than 0.05 was considered as statistically significant.

2.3. Results

2.3.1. Characterization of 3D hydrogels with controlled, anisotropic LN-1 concentrations

2.3.1.1. Generation of isotropic LN-1-agarose scaffolds

Increasing the amount of LN-1-SANPAH conjugate initially added to the agarose solution led to proportionally higher amounts of LN-1 being coupled to the agarose scaffold (Figure 2-4). When the amount of LN-1 initially added to agarose scaffold was doubled from 85 μg/ml to 170 μg/ml, the amount of LN-1 coupled to the scaffold increased more than three times (from 8.07 to 26.02 μg/ml) suggesting that the relationship between the amount of LN-1 added and amount of LN-1 coupled to scaffold is non-linear. The coupling efficiencies ranged from 9.5% (w/w) for the 85 μg/ml LN-1-agarose solutions to 15.3% (w/w) for the 170 μg/ml LN-1-agarose solutions (Figure 2-4). The coupling efficiencies for 133.33 μg/ml (13.41%) and 170 μg/ml (15.31%) were not significantly different (as compared by student's paired t-test). This suggests that the coupling efficiency had saturated at around 15% (w/w) for the range of LN-1 concentrations used in our study.

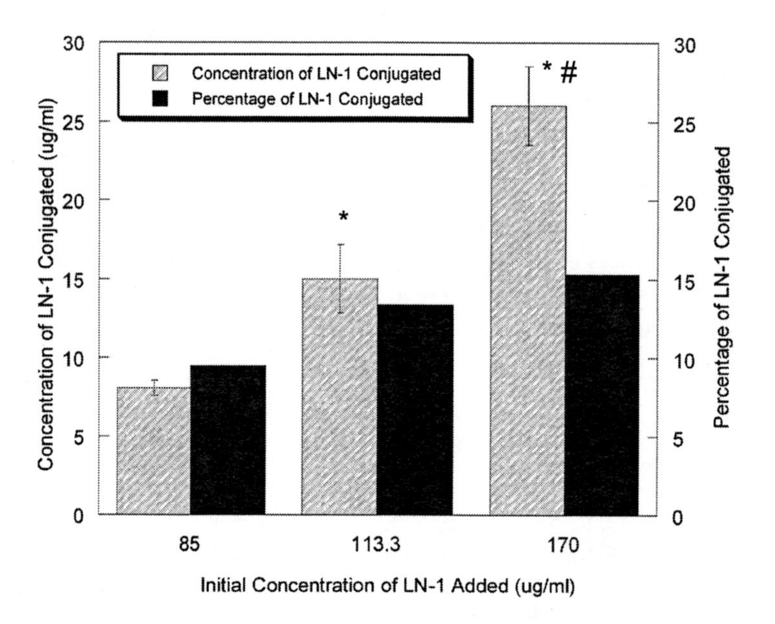

Figure 2-4: Photochemical immobilization of isotropic LN-1-concentrations in agarose scaffolds. Note that the concentration of LN-1 coupled to agarose scaffold increased when higher concentrations of LN-1 were added. The percentage of LN-1 coupled to agarose also increased with concentration but saturated around 15.31%. * represents $p < 0.05$ vs. 85 μg/ml data, # represents $p < 0.05$ vs. 133.33 μg/ml data; student's paired t-test was performed for n =3 samples/group.

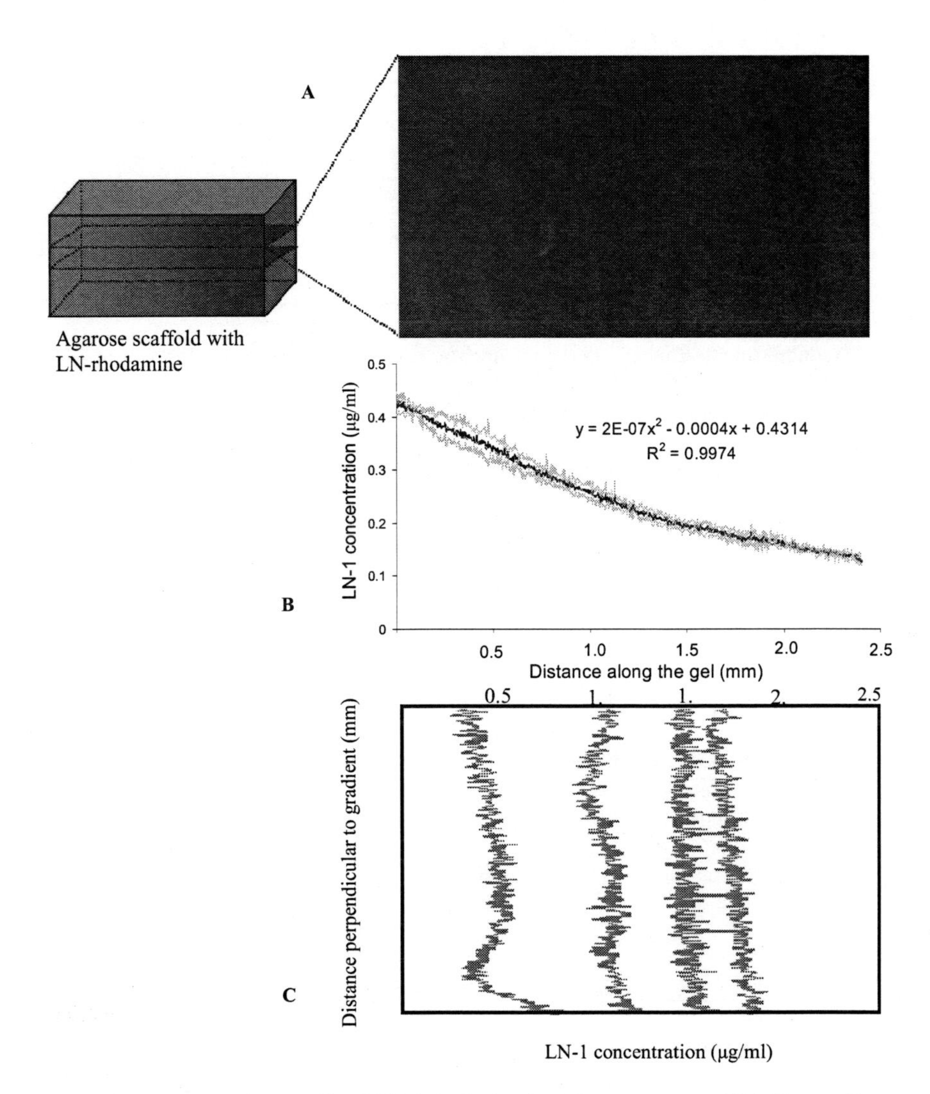

Figure 2-5: Concentration profile of LN-1-rhodamine bound to agarose scaffold. The gradient shown here was obtained when 282 µg/ml of LN-1 solution was added to the "high" compartment and PBS was added to the "low" compartment. The concentration profiles in the direction along (Figure 2-5B) and perpendicular (Figure 2-5C) to the gradient were obtained by averaging intensity profiles from 3 images at different depths from the same scaffold block. In Figure 2-5C, the dark line indicates the average concentration, gray lines indicate standard deviations from the average values, and the red line indicates the polynomial fit to the average concentration profile. A second-order polynomial fit gives a close approximation of the concentration distribution.

2.3.1.2. Characterization of LN-1 gradients in agarose scaffolds

Anisotropic LN-1 scaffolds with three different concentrations ranges, 0.430-0.126 μg/ml, 0.250- 0.123 μg/ml and 0.162- 0.120 μg/ml, having different slopes of LN-1 concentration gradients, 0.121 μg/ml/mm, 0.051 μg/ml/mm and 0.017 μg/ml/mm, respectively, were obtained by adding different concentrations of LN-1 in the "high" compartment. Figure 2-5 depicts the distribution of immobilized rhodamine-LN-1 across a 2.5 mm wide agarose scaffold block along the direction of gradient (Figure 2-5B) and also perpendicular to the gradient (Figure 2-5C). This LN-1 distribution was obtained when a 282 μg/ml LN-1 solution was added to the "high" concentration compartment of the scaffold and PBS solution was added to the "low" concentration compartment. Immobilization by UV light was done after 6 hours of diffusion. An average intensity profile was obtained by taking fluorescence intensity images at three different, fixed levels within the same agarose scaffold block and averaging the intensity values. The LN-1 distribution in the anisotropic scaffolds could be approximated by a second-order polynomial as shown in Figure 2-5B.

<u>2.3.2. Neurite growth from DRGs cultured in gradients of LN-1</u>

2.3.2.1. Neurite extension in isotropic LN-1 scaffolds

Three culture wells were used for DRG culture for each of the different isotropic LN-1 concentration scaffolds: 0, 0.2, 0.5, 0.8, 1.02, 3.4 and 10.21 μg/ml. For each condition, images of DRGs and their neurites were captured on day 1, day 2, day 3 and day 4. Around 30-35 neurites from each of 5-6 different DRGs, which could be reliably followed over the 4 days of culture, were chosen for growth rate measurements. The rate

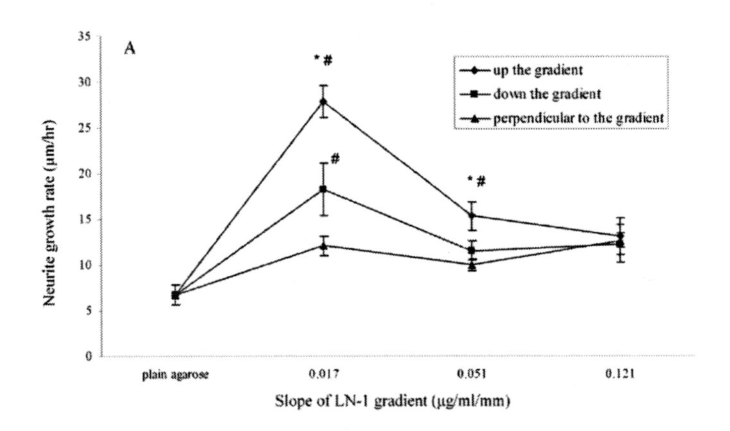

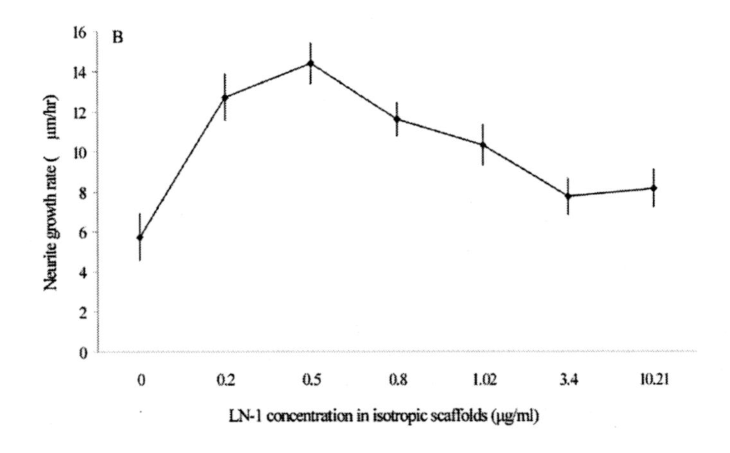

Figure 2-6: Neurite growth rate data from DRG explants in LN-1-agarose scaffolds. (A) Neurite growth rate in anisotropic LN-1 scaffolds. One-way ANOVA was performed for n = 30-35 neurites/group. * indicates p-value < 0.05 compared to neurites growing "down the gradient", # indicates p-values < 0.05 compared to neurites growing "perpendicular to the gradient". (B) Neurite growth rate in isotropic LN-1 scaffolds with different LN-1 concentrations. All error bars indicate standard error of mean (SEM).

of neurite extension was calculated by measuring neurite lengths at different time points. Care was taken to see that the length of the same neurite was measured at different time points by using fixed landmarks in the culture well. Maximum neurite growth rate was found for the LN-1 concentration range of 0.2-0.5 μg/ml. The rate decreased at higher LN-1 concentrations (Figure 2-6B).

2.3.2.2. Neurite extension in anisotropic LN-1 scaffolds

DRG neurite growth rate in three different LN-1 gradients: 0.121 μg/ml/mm (from 0.430- 0.126 μg/ml, over 2.5 mm thick agarose scaffolds), 0.051 μg/ml/mm (from 0.250 μg/ml - 0.123 μg/ml) and 0.017 μg/ml/mm (from 0.162 μg/ml - 0.120 μg/ml) (Figure 2-6A), was determined similar to isotropic LN-1 culture conditions. In all the three LN-1 gradients, similar initial concentration of LN-1 for the lower end was applied (around 0.120 μg/ml), so that the neurites were exposed to identical initial concentration of LN-1. The number of neurites and DRGs considered for rate measurements were identical to the isotropic LN-1 scaffold conditions. The neurite growth rate was highest for the lowest slope of LN-1 gradient, 0.017 μg/ml/mm. As the slope of the LN-1 gradient was increased, the neurite growth rate decreased.

For 0.017 μg/ml/mm gradient, neurites growing "up the gradient" grew significantly faster than neurites growing "down the gradient" ($p < 0.01$), which in turn grew faster than neurites growing "perpendicular to gradient" ($p < 0.05$). For 0.051 μg/ml/mm gradient, neurites grew "up the gradient" significantly faster than neurites growing "down the gradient" and neurites growing "perpendicular to the gradient", but the neurites growing "down the gradient" did not grow faster than neurites growing

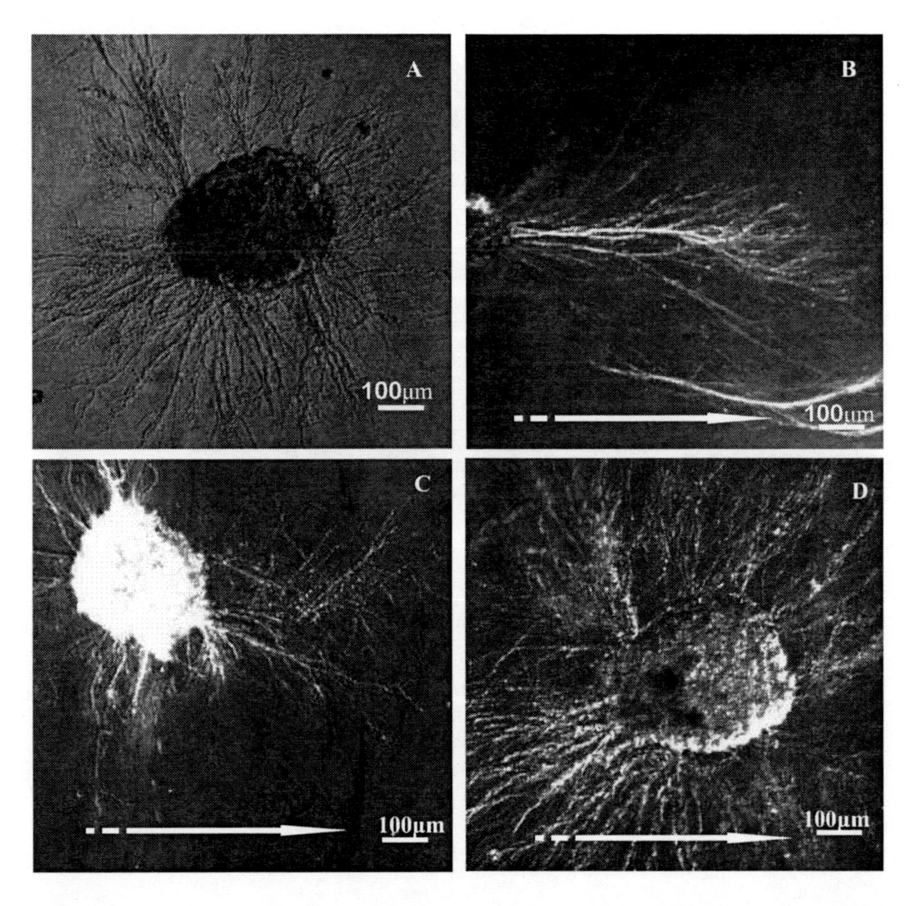

Figure 2-7: DRG cultured in anisotropic LN-1-agarose scaffolds. (A) plain agarose, (B) 0.017 μg/ml/mm (C) 0.051 μg/ml/mm and (D) 0.121 μg/ml/mm scaffold. The white arrow indicates the direction of increasing concentration of LN-1 (Scale bar = 100 μm). Plain agarose (A) and 0.121 μg/ml/mm scaffolds (D) promoted neurite extension uniformly in all directions, whereas 0.017 μg/ml/mm (C) 0.051 μg/ml/mm scaffolds promoted more neurite extension in the direction of gradients.

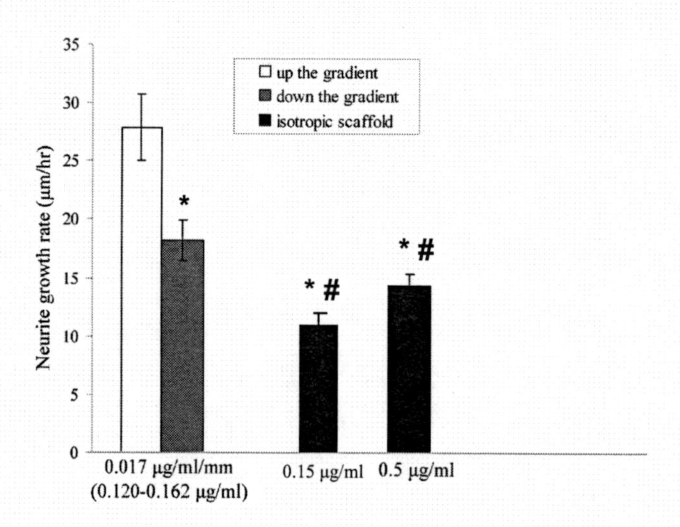

Figure 2-8: Comparison of neurite growth rate between anisotropic and isotropic LN-1 scaffolds using one-way ANOVA. * indicates p-value < 0.01 compared to "up the gradient" group, # indicates p-value < 0.01 compared to "down the gradient" group. The numbers in the bracket for 0.017 μg/ml/mm scaffold show the range of LN-1 concentration in the scaffold. Error bars indicate SEM.

"perpendicular to the gradient". For 0.121 μg/ml/mm gradient, there was no significant difference between growth rate of neurites growing "up the gradient", "down the gradient" and "perpendicular to the gradient". Representative phase-contrast microscopy images of DRG cultured in the anisotropic scaffolds and plain agarose have been shown in Figure 2-7.

The best neurite growth rate in anisotropic scaffolds was compared to the best neurite growth rate in isotropic scaffolds (Figure 2-8). Fastest neurite growth rate in anisotropic scaffolds was at a LN-1 gradient of 0.017 μg/ml/mm (0.120 μg/ml – 0.162 μg/ml) and in isotropic scaffolds was at 0.5 μg/ml LN-1 concentration. In anisotropic scaffold neurite extension "up the gradient" (27.8 μm/hr) and "down the gradient" (18.2 μm/hr) were significantly faster than neurite extension in 0.5 μg/ml isotropic LN-1 scaffold (14.38 μm/hr).

In order to test if LN-1 gradients orient the direction of neurite outgrowth, the total number of neurites growing "up the gradient", "down the gradient" and "perpendicular to the gradient" were counted for each culture condition. No significant differences between these numbers were observed for any of the conditions. It was also noticed that the neurite which had started growing "down the gradient" or "perpendicular to the gradient" did not try to reorient itself "up the gradient", but continued growing along the same direction.

2.4. Discussion

A technique to generate diffuse gradients of nerve growth factor (NGF) in 3D scaffolds has been previously described by Cao et al.[23] As a strategy to promote regeneration across long gaps, immobilized gradients may be more sustainable than diffuse gradients over the period of several weeks during which PNS regeneration occurs. This study presents a technique to fabricate anisotropic 3D scaffolds with immobilized gradients of LN-1 and quantitatively evaluates the influence of these scaffolds on chick

DRG neurite extension. Isotropic LN-1 scaffolds and plain agarose scaffolds without LN-1 were used as controls. The results demonstrate that particular concentration gradients of LN-1 promote faster neurite extension than the highest neurite growth rate observed with isotropic LN-1 concentrations. This suggests that gradients of LN-1 may be a more important parameter to optimize for maximal neurite extension rather than the absolute concentrations of isotropically distributed LN-1. The highest growth rate observed in anisotropic scaffolds was almost 5-times the growth rate in plain agarose gels. When DRGs were cultured in anisotropic LN-1 scaffolds with a steep slope of LN-1 gradient (0.121 μg/ml/mm or 241% concentration difference across 2.5 mm scaffold), there was no difference in neurite growth rate "up the gradient", "down the gradient" and "perpendicular to gradient". When the steepness of the gradient was reduced to 0.051 μg/ml/mm (103.3% concentration difference across 2.5 mm scaffold), there was difference between neurite growth rate "up the gradient" and "down the gradient" (p-value < 0.05). When the steepness of the gradient was further reduced to 0.017 μg/ml/mm (or 35% concentration difference across 2.5 mm scaffold), there was even more difference between neurite growth "up the gradient" and "down the gradient" (p-value < 0.01).

LN-1 is an appropriate candidate as the protein of choice for PNS regeneration since it has been shown to be a potent stimulator of neuronal cell migration and neurite outgrowth [13,124,126]. The spatio-temporally controlled expression of LN-1 in the developing peripheral nervous system, as well as the visual pathway and the cerebellum, suggests that LN plays a role in axon outgrowth and guidance [99,111,125]. LN expression has been found to be essential for ventral turning of axons during development. Blocking nidogen-binding motifs on LN prevents turning of axons although it does not affect growth [17], indicating that LN might play a role in orientation of axons in some systems.

Most previous studies have examined the effects of sharp boundaries of LN itself or LN with other molecules to direct neurite outgrowth [47,108] in 2D cultures. Even in other studies which have examined the directionality of axonal elongation on gradients of substratum bound molecules, 2D cultures have been used [1,66,110]. In our study we have used a 3D culture model with immobilized gradients of LN-1, which may be of greater physiological relevance as, *in vivo*, the axons have to navigate a 3D ECM environment, and the changes in LN concentrations are more likely to be gradual than sharp. Previous studies have examined the directionality of axonal elongation on gradients of substratum-bound molecules. McKenna et al. have shown that growth cones of sympathetic neurons show no directional response to substratum-bound laminin gradient [110]. On the other hand, Halfter has shown that elongation of retinal axons was affected by gradients of merosin (laminin-2) or a basal lamina extract [66]. Axons growing down a gradient were shorter and more fasciculated, while axonal turning was noted for axons growing up a shallow gradient. Adams et al. have reported that axonal growth cones can turn and migrate up a substratum-bound gradient of peptide from the LN-1 $\alpha1$ chain on 2D substrates [1]. But once the growth cones were oriented along the gradient, whether "up the gradient" or "down the gradient", they grew rapidly in either direction. In the anisotropic scaffolds used in our studies, the extending neurites were exposed to either increasing or decreasing gradient of LN-1. The neurites that were exposed to decreasing gradient did not try to reorient themselves "up the gradient", but continued to grow "down the gradient". Similarly, neurites exposed to increasing gradient of LN-1 did not try to reorient themselves "down the gradient". However, the steepness of the gradients influenced the rate of neurite extension, as described earlier in this section. In this study, the steepest gradient of 0.121 μg/ml/mm yielded a 2.41% concentration change across 25 μm, which is the mean diameter of a growth cone. In the 0.051 μg/ml/mm gradient, there was a 1.03% concentration change across the growth cone, and in the 0.017 μg/ml/mm gradient, there was a 0.35% concentration change across the growth cone. The percentage of change in concentration across the growth cone, at which we observed differences in neurite growth rates in these studies, was generally lower than that used by other studies (10% by Adams et al., and 3.5% by McKenna et al.). However, the amounts of LN-1 in gradients used in this study were generally higher compared to the above mentioned studies, which could have enhanced our ability to observe changes in the neurite growth rates. Interestingly, it was observed that when the steepness of LN-1 gradients (consequently the % change in concentration across growth cone) was decreased from 0.121 μg/ml/mm (2.41%) to 0.051 μg/ml/mm (1.03%) to 0.017 μg/ml/mm (0.35%), the neurites growth rate increased, and the rate of neurite growth "up the gradient" was faster than rate of neurite growth "down the gradient". This could be due to regulation of receptors for LN-1 by chick DRG. Condic et al. have shown that chick DRG down-regulate receptors for LN-1 when cultured in high concentrations of LN-1 and up-regulate the receptors when cultured in low concentrations of LN-1 [33]. It is possible that in the 0.017 μg/ml/mm LN-1 gradient scaffolds, DRGs express more receptors for LN-1 (at the growth cones) than in the 0.051 μg/ml/mm and the 0.121 μg/ml/mm gradient scaffolds. Substrate-cytoskeletal coupling in the growth cone could transmit mechanical forces to pull the growth cone or its contents toward the anchored adhesive site. Therefore, in the anisotropic LN-1 scaffolds with lower slopes, higher density of receptors might lead to faster neurite growth rates by haptotactic mechanism.

In summary, a novel technique to fabricate anisotropic LN-1 containing hydrogel scaffolds was fabricated and characterized. The results clearly demonstrate that scaffolds with gradients of LN-1 can promote faster neurite growth rate than maximal growth rate possible in isotropic LN-1 scaffolds. The results also demonstrate that there exists an optimal 'slope' for the gradient of LN-1, with the lowest slope in the range tested being the most optimal for the highest neurite growth rate. Although some studies have been done to understand how neural cells regulate their integrin expression in response to different isotropic concentrations of LN [33], in future studies it would be interesting to investigate how neurites adapt to changing concentrations of LN as in the case of gradients. Such anisotropic scaffolds, with directional cues, are likely to enhance peripheral nerve regeneration *in vivo* when presented across nerve gaps inside carrier polymer nerve guidance channels.

2.5. References

[1] Adams, D.N., Kao, E.Y., Hypolite, C.L., Distefano, M.D., Hu, W.S. and Letourneau, P.C., Growth cones turn and migrate up an immobilized gradient of the laminin IKVAV peptide, *J Neurobiol*, 62 (2005) 134-47.

[2] Baier, H. and Bonhoeffer, F., Axon guidance by gradients of a target-derived component, *Science*, 255 (1992) 472-5.

[3] Baron-van Evercooren, A., Kleinman, H.D., Ohno, S., Marangos, P., Schwartz, J.P. and Dubois-Dalcq, M.E., Nerv growth factor, laminin and fibronectin promote nerve growth in human fetal sensory ganglia cultures, *J Neurosci. Res.* (1982) 179-183.

[4] Bellamkonda, R.V., Ranieri, J.P., Bouche, N. and Aebischer, P., Hydrogel-based three-dimensional matrix for neural cells, *J Biomed Mater Res*, 29 (1995) 663-71.

[5] Bonhoeffer, F. and Huf, J., In vitro experiments on axon guidance demonstrating an anterior-posterior gradient on the tectum, *Embo J*, 1 (1982) 427-31.

[6] Bonner, J. and O'Connor, T.P., The permissive cue laminin is essential for growth cone turning in vivo, *J Neurosci*, 21 (2001) 9782-91.

[7] Bronner-Fraser, M., Stern, C.D. and Fraser, S., Analysis of neural crest cell lineage and migration, *J Craniofac Genet Dev Biol*, 11 (1991) 214-22.

[8] Cao, X. and Shoichet, M.S., Defining the concentration gradient of nerve growth factor for guided neurite outgrowth, *Neuroscience*, 103 (2001) 831-40.

[9] Champion, S., Imhof, B.A., Savagner, P. and Thiery, J.P., The embryonic thymus produces chemotactic peptides involved in the homing of hemopoietic precursors, *Cell*, 44 (1986) 781-90.

[10] Clark, P., Britland, S. and Connolly, P., Growth cone guidance and neuron morphology on micropatterned laminin surfaces, *J Cell Sci*, 105 (Pt 1) (1993) 203-12.

[11] Condic, M.L. and Letourneau, P.C., Ligand-induced changes in integrin expression regulate neuronal adhesion and neurite outgrowth, *Nature*, 389 (1997) 852-856.

[12] Dertinger, S.K., Jiang, X., Li, Z., Murthy, V.N. and Whitesides, G.M., Gradients of substrate-bound laminin orient axonal specification of neurons, *Proc Natl Acad Sci U S A*, 99 (2002) 12542-7.

[13] Dillon, G.P., Yu, X. and Bellamkonda, R.V., The polarity and magnitude of ambient charge influences three-dimensional neurite extension from DRGs, *J Biomed Mater Res*, 51 (2000) 510-9.

[14] Dillon, G.P., Yu, X., Sridharan, A., Ranieri, J.P. and Bellamkonda, R.V., The influence of physical structure and charge on neurite extension in a 3D hydrogel scaffold, *J Biomater Sci Polym Ed*, 9 (1998) 1049-69.

[15] Esch, T., Lemmon, V. and Banker, G., Local presentation of substrate molecules directs axon specification by cultured hippocampal neurons, *J Neurosci*, 19 (1999) 6417-6426.

[16] Garcia-Alonso, L., Fetter, R.D. and Goodman, C.S., Genetic analysis of Laminin A in Drosophila: extracellular matrix containing laminin A is required for ocellar axon pathfinding, *Development*, 122 (1996) 2611-21.

[17] Halfter, W., The behavior of optic axons on substrate gradients of retinal basal lamina proteins and merosin, *J Neurosci*, 16 (1996) 4389-401.

[18] Kendal, E., Schwartz, J. and Jessal, T.M., The induction and patterning of nervous system., *Principles of Neuroscience*, McGraw-Hill, 2000, pp. 1019-1040.

[19] Letourneau, P.C., Chemotactic response of nerve fiber elongation to nerve growth factor, *Dev Biol*, 66 (1978) 183-96.

[20] Liesi, P., Do neurons in the vertebrate CNS migrate on laminin?, *Embo J*, 4 (1985) 1163-70.

[21] Liesi, P., Hager, G., Dodt, H.U., Seppala, I. and Zieglgansberger, W., Domain-specific antibodies against the B2 chain of laminin inhibit neuronal migration in the neonatal rat cerebellum, *J Neurosci Res*, 40 (1995) 199-206.

[22] Matsumoto, K., Ohnishi, K., Kiyotani, T., Sekine, T., Ueda, H., Nakamura, T., Endo, K. and Shimizu, Y., Peripheral nerve regeneration across an 80-mm gap bridged by a polyglycolic acid (PGA)-collagen tube filled with laminin-coated collagen fibers: a histological and electrophysiological evaluation of regenerated nerves, *Brain Res*, 868 (2000) 315-28.

[23] Matsuzawa, M., Tokumitsu, S., Knoll, W. and Leisi, P., Molecular gradeint along the axon pathway is not required for directional axon growth, *J Neurosci Res*, 53 (1998) 114-124.

[24] McKenna, M.P. and Raper, J.A., Growth cone behavior on gradients of substratum bound laminin, *Dev Biol*, 130 (1988) 232-236.

[25] McLoon, S.C., McLoon, L.K., Palm, S.L. and Furcht, L.T., Transient expression of laminin in the optic nerve of the developing rat, *J Neurosci*, 8 (1988) 1981-90.

[26] Powell, S.K. and Kleinman, H.K., Neuronal laminins and their cellular receptors, *Int J Biochem Cell Biol*, 29 (1997) 401-14.

[27] Rivas, R.J., Burmeister, D.W. and Goldberg, D.J., Rapid effects of laminin on the growth cone, *Neuron*, 8 (1992) 107-115.

[28] Rogers, S.L., Edson, K.J., Letourneau, P.C. and McLoon, S.C., Distribution of laminin in the developing peripheral nervous system of the chick, *Dev Biol*, 113 (1986) 429-35.

[29] Rogers, S.L., Letourneau, P.C., Palm, S.L., McCarthy, J. and Furcht, L.T., Neurite extension by central and peripheral nervous system neurons in response to substratun-bound fibronectin and laminin, *Dev Biol*, 98 (1983) 212-220.

[30] Schmidt, C.E. and Leach, J.B., Neural Tissue Engineering: Strategies for repair and regeneration, *Annu. Rev. Biomed. Eng* (2003) 293-347.

[31] Song, H. and Poo, M., The cell biology of neuronal navigation, *Nat Cell Biol*, 3 (2001) E81-8.

[32] Tessier-Lavigne, M. and Goodman, C.S., The molecular biology of axon guidance, *Science*, 274 (1996) 1123-33.

[33] Uckun, F.M., Evans, W.E., Forsyth, C.J., Waddick, K.G., Ahlgren, L.T., Chelstrom, L.M., Burkhardt, A., Bolen, J. and Myers, D.E., Biotherapy of B-cell precursor leukemia by targeting genistein to CD19-associated tyrosine kinases, *Science*, 267 (1995) 886-91.

[34] Valentini, R.F. and Aebischer, P., Strategies for the engineering of peripheral nervous tissue regeneration. In L.R. Lanza RP, Chick WL (Ed.), *Principles of Tissue Engineering*, R.G.Landes Company, Austin, 1997, pp. 671-684.

[35] Yu, X. and Bellamkonda, R.V., Tissue-engineered scaffolds are effective alternatives to autografts for bridging peripheral nerve gaps, *Tissue Eng*, 9 (2003) 421-30.

[36] Yu, X., Dillon, G.P. and Bellamkonda, R.B., A laminin and nerve growth factor-laden three-dimensional scaffold for enhanced neurite extension, *Tissue Eng*, 5 (1999) 291-304.

[37] Zhou, F.C. and Azmitia, E.C., Laminin directs and facilitates migration and fiber growth of transplanted serotonin and norepinephrine neurons in adult brain, *Prog Brain Res*, 78 (1988) 413-26.

Chapter 3. **ANISOTROPIC AGAROSE HYDROGEL SCAFFOLDS ENHANCE NERVE REGENERATION ACROSS LONG NERVE GAPS IN RODENTS**

This chapter is in preparation as a manuscript.

Tubular Polysulfone guidance channels filled with anisotropic scaffolds, agarose hydrogels containing gradients of laminin-1 (LN-1) and nerve growth factor (NGF), were implanted to promote sciatic nerve regeneration across a 20-mm nerve gap in rats. Isotropic scaffolds with uniform concentration of LN-1 and NGF were used as negative control, and sciatic nerve grafts harvested from syngenic rats were used as positive control. 4-months after implantation, regenerating axons were found only in the scaffolds having gradients of both LN-1 and NGF molecules. Anisotropic scaffolds with gradient of only LN-1 or only NGF, with the other component at uniform concentration, did not result in axonal formation as observed by histological analysis. This suggests that LN-1 and NGF act synergistically to enhance nerve regeneration. Nerve grafts also resulted in axonal regeneration across the 20-mm nerve gap. The total number of myelinated axons was similar for anisotropic scaffolds and nerve grafts, with the anisotropic scaffolds having higher density of axons than nerve grafts. Axonal diameter distribution was similar for anisotropic scaffolds and nerve grafts. Nerve grafts, however, resulted in better functional outcome as measured by relative gastrocnemius muscle weight (RGMW) and electrophysiology. These results indicate that for nerve regeneration over long-nerve gaps, anisotropic scaffolds perform better than isotropic scaffolds. However, the concentrations of LN-1 and NGF in the anisotropic scaffolds need to be optimized to perform better than nerve grafts.

3.1. Introduction

Severe traumatic nerve injuries, tumor removal and congenital anomalies may result in injuries to critical nerves, and formation of a nerve gap. Failure to repair the nerve gaps can result in the loss of muscle function, impaired sensation and/or painful neuropathies. In case of a small nerve gap (< 2-mm long), the injury can be repaired by approximating the two injured nerve ends and suturing/gluing them together [86,140]. When the nerve gap is longer than 2-mm, as in most clinical conditions, the standard technique is to transplant autologous nerve graft from an uninjured site to the injured site. Even though nerve grafts are considered as the 'gold standard' for nerve repair, they do not result in complete nerve regeneration and recovery [114,166]. In addition, an autologous nerve graft also suffers from the drawbacks that it requires sacrifice of a healthy functional tissue, and additional surgery time. To avoid the problems with autografts, allografts from cadavers have been used. However, allografts usually require preservation between surgeries and systemic immunosuppression in recipients, both of which can reduce allograft performance [62,115].

To overcome the limitations with natural materials, synthetic tubular nerve guidance channels have been used with promising results [6,10,26,151]. The performance of nerve guidance channels (NGCs) can be further enhanced by varying its properties, such as, porosity [2], electrical activity [5,153], and surface roughness [3]. In the presence of distal nerve stump, silicone NGCs can promote spontaneous nerve regeneration over 10-mm nerve gaps in rats. However, regeneration fails if the nerve gap is increased to 15-mm or longer [36,103]. To enhance the performance of NGCs, additional filler materials (such as hydrogels), growth factors, extracellular matrix (ECM) proteins, and fibers have been used [46,105,107,147].

The ECM protein laminin (LN), found in the basement membrane of most cells, is a potent stimulator of neurite outgrowth in neurons from peripheral and central nervous systems. Also, LN enhances attachment and migration of neuronal cells and increases survival of neurons [84,101,122,124]. Nerve growth factor (NGF) is a neurotrophic factor produced by the target organs of sympathetic and sensory nerves [12]. NGF has been shown to stimulate and promote neurite outgrowth and the survival of sensory neurons and axons, including those from spinal sensory nerves and sciatic nerves [73,95,145]. Due to their potent functions, LN and NGF have been used successfully for nerve regeneration, both *in vitro* and *in vivo* [7,13,24,38,164].

Various hydrogel scaffolds, including agarose, have been used to promote neuronal growth, both *in vitro* and *in vivo* [130,141,159]. Agarose hydrogel is a polysaccharide derived from red algae. In cell cultures, SeaPrep® agarose has been shown to support neurite extension from a variety of neurons in a non-immunogenic manner [15,40,89,137]. Agarose gels also allow molecules to be covalently linked to the gels through functional groups on its polysaccharide chains. For example, LN-1 protein (isoform of LN) or fragments of LN-1 can be covalently coupled to SeaPrep® agarose gels to further enhance their ability to support neurite extension [164]. Photochemical conjugation can be used to photoimmobilize gradients of LN-1 in agarose gels to promote enhanced and directed neurite extension *in vitro* [42].

Neurotrophic factors used to promote nerve regeneration have short half-lives, and hence, in case of regeneration over long nerve gaps, ideally need to be slow-released using delivery vehicles. For slow-release of neurotrophic factors, phosphatidyl choline based lipid microtubules have been developed [127,136]. Lipid microtubules (LMTs) are ideal for trophic factors because these sensitive proteins can be loaded into microtubules in an aqueous environment without exposure to organic solvents. LMTs can also be easily incorporated in agarose scaffolds, when culturing neuronal cells, for slow-release, without physically impeding the growth cones that navigate the scaffolds. By controlling the amount of protein loaded into the microtubules, the number of microtubules used, and the length of microtubules; it is possible to control the duration of release of protein and the amount of protein released [113].

Tissue-engineered scaffolds have been successful in promoting nerve regeneration comparable to nerve grafts, across short nerve gaps (10-mm or less). Polymer guidance channels with uniform concentrations (*isotropic*) of LN-1 and NGF promote nerve regeneration comparable to that of autografts over 10 mm nerve gaps in rodents [163]. However, we believe that for longer nerve gaps, 15-mm or more in rodents or over 80 mm in humans, which is clinically more relevant, *isotropic* scaffolds might not be able to match the performance of autogratfts. *Anisotropic* agarose hydrogel scaffolds, with gradients of LN-1, have been shown to promote enhanced neurite extension from chick dorsal root ganglia (DRG) *in vitro* as compared to *isotropic* scaffolds [42]. DRG cultured in these *anisotropic* scaffolds extend neurites longer and faster than DRG cultured in *isotropic* LN-1 scaffolds. Similarly, gradients of NGF can initiate turning of neurites towards the NGF-source [52], and guide the neurite growth [23,24]. We hypothesize that *anisotropic* scaffolds with gradients of both LN-1 and NGF will promote better nerve regeneration than *isotropic* scaffolds over large nerve gaps. Here, we present our techniques to synthesize *anisotropic* scaffolds using agarose gels, the animal model we have used, analyses for nerve regeneration and behavior studies, and the results.

3.2. Materials and Methods

3.2.1. Design of Tissue-Engineered Scaffolds

3.2.1.1. Preparation of Polymer Guidance Channel

Tubular Polysulfone guidance channels were used to present the LN-1 and NGF-containing agarose hydrogel scaffolds, for sciatic nerve regeneration in rats. Polysulfone guidance channels (Koch Membrane System, Ann Arbor, MI), with an inner diameter of 1.6 mm, an outer diameter of 3.2 mm, and length of 22 mm, were used for supporting the agarose gel scaffolds. Before filling the guidance channels with the agarose scaffold, they were sterilized by immersion in 70% ethanol solution for 2 days, dried under a clean laminar flow hood, and washed with sterilized Phosphate Buffered Saline (PBS) (Mediatech Inc., Herndon, VA). The guidance channels were kept hydrated in 0.1M PBS (pH 7.4) until they were filled with agarose scaffolds.

3.2.1.2. Synthesis of LN-1 coupled Agarose Hydrogels and Lipid Microtubules for Slow Release of NGF

Agarose gel scaffolds were designed with three components: 0.5% (w/v) agarose hydrogel, LN-1, and NGF-loaded microtubules; for spatially and temporally controlled presentation of ECM and trophic factors *in vivo*. First, thermo-reversible SeaPrep® agarose hydrogels (BMA, Rockland, Rockland, ME) were covalently coupled with ECM protein LN-1 using photochemical conjugation [42]. Briefly, LN-1 (BD Biosciences,

Bedford, MA) was first conjugated to a bi-functional photochemical crosslinker, Sulfo-SANPAH (Sulfosuccinimidyl-6-[4'-azido-2'-nitrophenylamino] hexanoate) (PIERCE, Rockford, IL) through the amine groups on the LN-1 molecule. Agarose solution was then added to this LN-1 – Sulfo-SANPAH conjugate such that final concentration of agarose is 1% (w/v). The solution mixture was then exposed to ultra-violet light, binding LN-1 to agarose through the photocrosslinker. The solution was then solidified into gel by cooling in a refrigerator for 20 minutes. Over the next 2 days, the gel was washed using 0.1M PBS, with repeated changes, to remove uncoupled LN-1. LN-1 conjugated agarose hydrogel was then liquefied by heating at 45°C, cooled to room temperature, mixed with NGF-loaded microtubules and injected into polymer guidance channels. Amount of LN-1 conjugated to agarose hydrogel was quantified by Bradford protein assay (BIO-RAD, Hercules, CA).

A novel drug delivery system using lipid microtubules (LMTs) of 1,2-bis (tricosa-10,12-diyomoyl)-sn-3-phosphocholine (DC8,9PC; Avanti Polar Lipids, Inc., Alabster, AL) was prepared by ethanol deposition method, and used for slow-release of NGF, as described in other studies [113,164]. The LMTs had an average length of 45±20 μm (Figure 3-1). NGF was loaded into the LMTs by hydration of 10 mg of lyophilized LMTs with 400 μl of PBS containing 48 μg of NGF (120 μg/ml NGF solution). The NGF loaded LMTs were then mixed with equal volume of 1% plain agarose or LN-1 modified agarose. NGF is slowly released by diffusion from the two open ends of the LMTs.

Figure 3-1: Lipid microtubules. Phase- contrast image of lipid microtubules from $DC_{8,9}PC$ lipid. Scale bar = 100 μm.

3.2.1.3. NGF release from NGF-loaded Microtubules

A study was designed to assess the long-term release of NGF from LMTs. LMTs were first synthesized by ethanol deposition method [113]. A total of 10 mg of LMTs was rehydrated with NGF solution (0.4 ml of 120 μg/ml solution) overnight at 4°C. The solution was then centrifuged to precipitate LMTs, and the supernatant containing free NGF was removed. The LMTs were mixed with equal volume of 2% agarose solution (w/v) to form a 1% (w/v) LMTs-embedded agarose solution. 300 μl of the mixture was added to a 24 well-plate dish (Corning Inc., Corning, NY) and cooled in refrigerator (4°C) for 20 minutes, to allow it to gel. 500 μl of PBS was added on top of the gel block and the 24 well-plate dish was maintained at 37°C. NGF slowly released from the two open ends of LMTs diffused into the PBS solution. The solution was replaced with fresh PBS daily, and the amount of NGF in the supernatant was quantified using an NGF ELISA Kit (Chemicon International, Temecula, CA) to determine the amount and duration of NGF-release from the LMTs.

3.2.1.4. Synthesis of Isotropic Scaffolds

Isotropic scaffolds with uniform concentration of LN-1 and NGF were designed as described below. LN-1 conjugated agarose solution (66 μg of LN-1/ml of 1% agarose) was synthesized as described in section 3.2.1.2., and mixed with an equal volume of LMTs in PBS solution (9.6×10^8 LMTs/ ml of PBS, loaded with 120 μg/ml of NGF). This resulted in 0.5% agarose solution with LN-1 (33 μg/ml) and NGF-loaded microtubules. The agarose solution mixture was then injected into Polysulfone guidance channels using a 1 ml syringe fitted with 22G needles (Becton Dickinson & Co., Franklin Lakes, NJ) and gelled by cooling at 4°C for 10 minutes. These scaffolds were kept hydrated in 0.1M PBS until implantation in rats, on the same day.

Two kinds of anisotropic scaffolds were designed, one with step-gradients, and the other with continuous-gradients. Both, step-gradient and continuous-gradient scaffolds had four layers of gels from one end of the tube to the other, each with higher concentration of NGF than the previous layer. However, in a step-gradient scaffold the LN-1 concentration increased in step-wise manner from one layer to another (Figure 3-2), whereas, in a continuous-gradient scaffold the LN-1 concentration increased smoothly from one end of tube to the other end (Figure 3-3).

3.2.1.5. Synthesis of Step-gradient Anisotropic Scaffolds

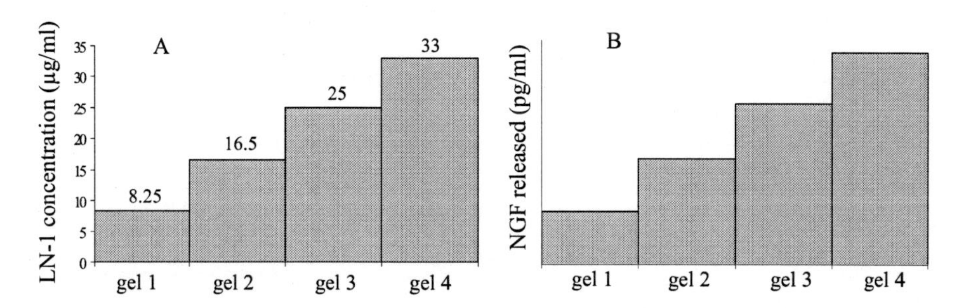

Figure 3-2: Schematic of LN-1 and NGF distribution in step-gradient anisotropic scaffolds. Gel 4 has higher concentration of LN-1 than gel 3 and so on (A). Similarly gel 4 has higher concentration of NGF than gel 3 (B) and so on. While LN-1 gradient is immobilized, with time, NGF will diffuse and form a smooth gradient.

The NGF-loaded LMTs and 1% (w/v) LN-1-agarose solution (66 µg of LN-1/ml) were mixed at 1:1 ratio to prepare 0.5% (w/v) agarose solution with immobilized LN-1 (33 µg/ml) and mixed in NGF-loaded LMTs (solution 4). Solution 4 was diluted with 0.5% agarose solution to form solution 3 with lower concentration of LN-1 and NGF-loaded LMTs. Similarly, solutions 2 & 1 were made. Solution 4 was first injected into the polymer guidance channel to fill 25% of space and allowed to gel. The solutions 3, 2 and 1, were then injected into the polymer guidance channel, one after the other, to make four layers of gels. The four layers resulted in increasing concentrations of LN-1 and NGF from one end of tube to the other end (Figure 3-2).

3.2.1.6. Synthesis of Continuous-Gradient Anisotropic Scaffolds

To design anisotropic scaffolds with continuous-gradients of LN-1 and NGF, first step-gradient of NGF was synthesized and a LN-1 gradient was made later. A 0.5% agarose solution mixture (solution 4) was first made by mixing 1% agarose solution (no LN-1) with equal volume of NGF-loaded LMTs (9.6 x 10^8 LMTs/ml of PBS). Solution 4 was diluted with 0.5% plain agarose solution to make Solution 3, with lower concentration of NGF-loaded LMTs. Similarly, solution 2 & 1 were made. Solution 4 was first injected into the polymer guidance channel to fill 25% of space, and allowed to gel by cooling. The solutions 3-1 were then injected into the polymer guidance channel, one after the other; to make four layers of gels, with increasing concentrations of NGF-loaded LMTs (no LN-1). A LN-1 gradient was then made by allowing controlled diffusion of LN-1-sulfo-SANPAH solution into the guidance channel through one of its ends. The LN-1 gradient was then immobilized by UV-photocrosslinking. Figure 3-3 shows a schematic of distribution of LN-1 (Figure 3-3A) and NGF-loaded LMTs (Figure 3-3B) in these scaffolds. To make *anisotropic* scaffolds with gradient of LN-1 but uniform NGF concentration, solutions 4-1 with same concentration of NGF-loaded LMTs were used, and then a diffusion gradient of LN-1 was made.

To determine the concentration profile of LN-1 in continuous-gradient scaffolds, the tubes were cut transversely into 4 parts, 5-mm each. The total amount of LN-1 in each section was determined by LN-1 ELISA.

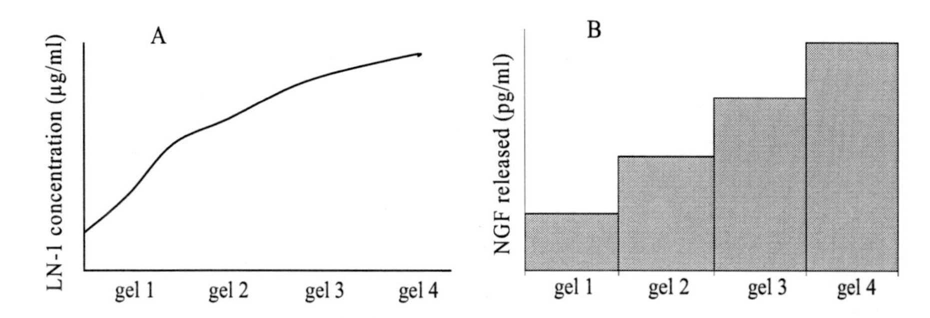

Figure 3-3: Schematic of LN-1 and NGF distribution in continuous-gradient anisotropic scaffolds. LN-1 gradient is smooth (A) while NGF is distributed in a step-gradient fashion initially (B). With time, NGF will diffuse and form a smooth gradient.

3.2.1.7. Different Groups of nerve implants

Nerve guidance channel (NGC) implants were separated into eight groups depending on their constituents, as described in Table 3-1. Control group I (Saline) contained 1X PBS solution. Control group II (Plain agarose) contained 0.5% agarose solution without any LN-1. Control group II (LN(U)) contained channels filled with LN-1-coupled 0.5% agarose gel (33 μg/ml of LN-1). Control group IV (nerve graft) consisted of 20-mm long nerve grafts harvested from isogenic Fisher inbred rats. Experimental group I (LN(U)-NGF(U)) consisted of implants with uniform concentration of LN-1 and NGF. Experimental group II (LN(U)-NGF(SG)) consisted of implants with uniform concentration of LN-1 and step-gradient of NGF. Experimental group III (LN(SG)-NGF(U)) consisted of implants with step-gradient of LN-1 and uniform concentration of NGF. Experimental group IV (LN(SG)-NGF(SG)) consisted of step-gradients of both LN-1 and NGF.

Table 3-1: Description of experimental and control groups for *in vivo* study			
Groups	**Notation of groups**	**Components**	**# of rats**
Control I (negative)	Saline	PBS solution	5
Control II (negative)	Plain agarose	0.5% agarose gels (no LN-1)	5
Control III (negative)	LN(U)	*Isotropic* LN-1-agarose gels	5
Control IV (positive)	Nerve graft	Nerve grafts explanted from other Fischer inbred rats	8
Experimental I	LN(U)-NGF(U)	*Isotropic* LN-1-agarose gels embedded with *isotropic* NGF-loaded LMTs	10
Experimental II	LN(U)-NGF(SG)	*Isotropic* LN-1-agarose gels embedded with *anisotropic* NGF-loaded LMTs	9
Experimental III	LN(SG)-NGF(U)	*Anisotropic* LN-1-agarose gels embedded with *isotropic* NGF-loaded LMTs	11
Experimental IV	LN(SG)-NGF(SG)	*Anisotropic* LN-1-agarose gels embedded with *anisotropic* NGF-loaded LMTs	9
Experimental V	LN(CG)-NGF(U)	*Anisotropic* LN-1-agarose gels embedded with *isotropic* NGF-loaded LMTs	7
Experimental VI	LN(CG)-NGF(SG)	*Anisotropic* LN-1-agarose gels embedded with *anisotropic* NGF-loaded LMTs	8

(U) = uniform concentration, (SG) = step-gradient, (CG) = continuous-gradient

Experimental group V (LN(CG)-NGF(U)) consisted of continuous gradient of LN-1 and uniform concentration of NGF. Experimental group VI (LN(CG)-NGF(SG)) consisted of implants with continuous gradient of LN-1 and step-gradient of NGF. After preparation, these implants were stored at 4°C in 0.1M PBS until implantation *in vivo*, usually within 2-3 hours.

3.2.2. Implantation of Tissue Engineered Scaffolds

We have used rat sciatic nerve injury model to test for nerve regeneration. Adult Fischer inbred male rats (Harlan) weighing between 300-350g were used. The rats were anesthetized using inhaled isofluorane gas (3-4% v/v for induction, 1.5-2% for maintenance, Airco, Inc., Madison, Wis.). The right thigh region was shaved and the skin was sterilized by applying chlorohexidine solution (2%, First Priority, Inc., Elgin, IL) and 70% ethanol alternatively, 2-3 times. A 25-mm long skin incision was made along the femoral axis. The thigh muscles were separated and the sciatic nerve was dissected free. Using micro-scissors, the nerve was transected and a 5-mm nerve segment was explanted. The proximal and distal nerve stumps were secured 20-mm apart in a 22-mm long polymer guidance channel carrying experimental or control gel formulations, using a 10-0 Nylon monofilament suture (Ethicon Inc., Somerville, NJ). In case of positive control group, the nerve gap was bridged using a 20-mm long syngenic nerve graft obtained from another Fischer inbred rat. The muscles were then closed using a 4-0 vicryl suture (Ethicon Inc., Somerville, NJ) and the skin was closed using wound clips (Braintree Scientific, Inc., Braintree, MA). 0.2 ml of Marcaine (0.25% w/v, Hospira, Inc., Lake Forest, IL) per rat was administered subcutaneously for pain relief. NIH guidelines for using experimental animals were followed both pre- and post-surgery. The explanted 5-mm nerve was fixed in para-formaldehyde and prepared for histological analysis to evaluate the native nerve prior to injury.

In case of *anisotropic* scaffolds the tubes were sutured to the nerve such that the concentration of LN-1 and NGF-releasing LMTs increases from the proximal end to the distal end. The rats were under observation for 16 weeks.

3.2.3. Evaluation of Nerve Regeneration

3.2.3.1. Histological Analysis for Nerve Regeneration

16 weeks post-implantation, after electrophysiological examinations, the rats were administered, intraperitoneally, an overdose of rodent anesthetic cocktail (consisting of ketamine at 65mg/kg of rat weight, xylazine at 7.5mg/kg, and acepromazine 0.5mg/kg). The rats were then perfused intracardially with saline, followed by cold 4% parafomaldehyde and 0.25% gluteraldehyde (both from Sigma-Aldrich, St. Louis, MO) in PBS. The site of nerve injury was opened and the implant (polymer guidance channel or nerve graft) was removed for histological analysis. Along with the nerve implant, the gastocnemius muscle from the right (experimental side) and left (control side) limb were also explanted. All the harvested tissues were post-fixed in 4% parafomaldehyde, overnight.

The nerve explants were cut into three parts: proximal, middle and distal, and additionally post-fixed, overnight, with 1% osmium tetroxide in PBS. After washing with PBS and dehydration in graded ethanol series, the three parts were separately embedded in LX112 resin (Ladd Research Industries, Inc., Burlington, VT). Semi-thin sections (0.5 μm) of nerve explants were cut using microtome, stained with Toluidine blue (0.1%, Sigma, St. Louis, MO) and observed under a light microscope. Ultra-thin sections (100 nm) were observed under electron microscope.

Nerve regeneration was evaluated by: a) percentage of guidance channels with successful nerve regeneration, indicated by presence of myelinated axons, b) the total number of myelinated axons in the middle (at 10-mm) and the distal end (at 17-mm) of the guidance channels/autografts, c) the area of axonal regeneration, d) the number of myelinated axons per unit area (density) at the middle and distal points of the guidance channel/autografts; and e) the diameter distribution of regenerated axons for each group. For quantification, images were captured using Sony digital photo camera (Japan) attached to Nikon Eclipse TE 300 microscope (Japan) running ImagePro software (Media Cybernetics, L.P., Silver Spring, MD). First, images were captured using 4x or 10x objective lenses to determine the area of axonal regeneration, and then images were captured using 100x objective lens to count the number of myelinated axons. 4-5 images (at 100x) were captured for each nerve implant, so that 15-50% of area of regeneration was used for analysis. One-way ANOVA was used for statistical comparison of the various groups.

3.2.3.2. Relative Gastrocnemius Muscle Weight Measurement

The gastrocnemius muscle is innervated by the sciatic nerve, and starts atrophying after nerve transection injury. The gastrocnemius muscle from the right and left limb were harvested after the rat had been sacrificed. The weight of the muscle was measured and used to calculate relative gastrocnemius muscle weight (RGMW). The RGMW, which is defined as the ratio of muscle from the experimental side to the control side, was used as one of the parameter for motor function recovery. The RGMW should increase with the sciatic nerve regeneration.

3.2.3.3. Neuromuscular junction (end plates), synaptic vesicles and neurofilament staining

Gastrocnemius muscle explanted from the experimental rats were post-fixed overnight in 4% paraformaldehyde, washed with saline solution and left overnight in 30% sucrose solution. The muscle tissue was then cryoembedded. Longitudinal sections, 25 μm thick, were cut using a Microtome (Cryo-star HM 560MV, Microm, Waldorf, Germany). Tissue sections were collected every 800 μm, and processed for the following markers: *neurofilament 160* (NF160, Sigma-Aldrich, St. Louis, MO), *synaptic vesicles 2 protein* (SV2, Developmental Studies Hybridoma Bank, Iowa City, IA), and *acetylcholine receptors* (using alpha-Bungarotoxin-tetramethlyrhodamine, Sigma-Aldrich, St. Louis, MO).

3.2.3.4. Electrophysiological Recordings to Test for Nerve Reconnection

To test for regeneration of axons through the implant and functionality of the regenerated nerve, compound action potentials (CAPs) across the implant were measured. Sixteen weeks post-implantation, prior to perfusing the rat for implant retrieval, electrophysiological recordings were done under anesthesia. The experimental rats were anesthetized using 0.7ml of rodent cocktail/kg of rat weight (consisting of ketamine at 65mg/kg of rat weight, xylazine at 7.5mg/kg, and acepromazine 0.5mg/kg), administered intraperotineally. The sciatic nerve, posterior tibial nerve and triceps surae muscles were exposed on the right limb. The sciatic nerve was stimulated 5-mm proximal to the implant and whole nerve CAPs were recorded 40-mm distal to the implant. The nerve was stimulated using a platform electrode (1 Hz, 3 ms, variable amplitude). Recordings were amplified 10,000 times, filtered at 10 Hz (low pass) and 3 kHz (high pass), acquired using LabVIEW software, and stored for offline averaging and analysis using MATLAB software. CAPs were also measured from the control side (left sciatic nerve) for 4 rats and denoted as "normal" CAPs.

3.2.3.5. Walking Track Analysis for Evaluation of Functional Recovery

Free, spontaneous locomotion in animals is highly consistent and readily quantifiable. In 1982, De Medinaceli et al. [37] designed a quantitative method of analyzing the sciatic nerve function in rats, known as the sciatic function index (SFI).

A walking track was designed to visualize and record gait of rats in an enclosed walkway (Figure 3-4). Each experimental rat was allowed to walk freely in the walkway and its movements were recorded using a digital camera. The video sequences of rat gait were edited in QuickTime Player (Apple Computer, Inc.) to capture the foot prints. The foot prints were then analyzed to measure (i) distance from the heel to the third toe, called as the print length (PL); (ii) distance from the first to the fifth toe, the toe spread (TS); and (iii) distance from the second to the fourth toe, the intermediate toe spread (ITS). All these measurements were taken from the experimental (E) and normal (N) sides. These measurements were used to calculate the factors as follows: (i) print length factor (PLF) = (EPL-NPL)/NPL; (ii) toe spread factor (TSF) = (ETS-NTS)/NTS; (iii) intermediate toe spread factor (ITF) = (EIT-NIT)/NIT. These factors were then incorporated into the Bain-Mackinnon-Hunter (BMH) sciatic function index-formula:

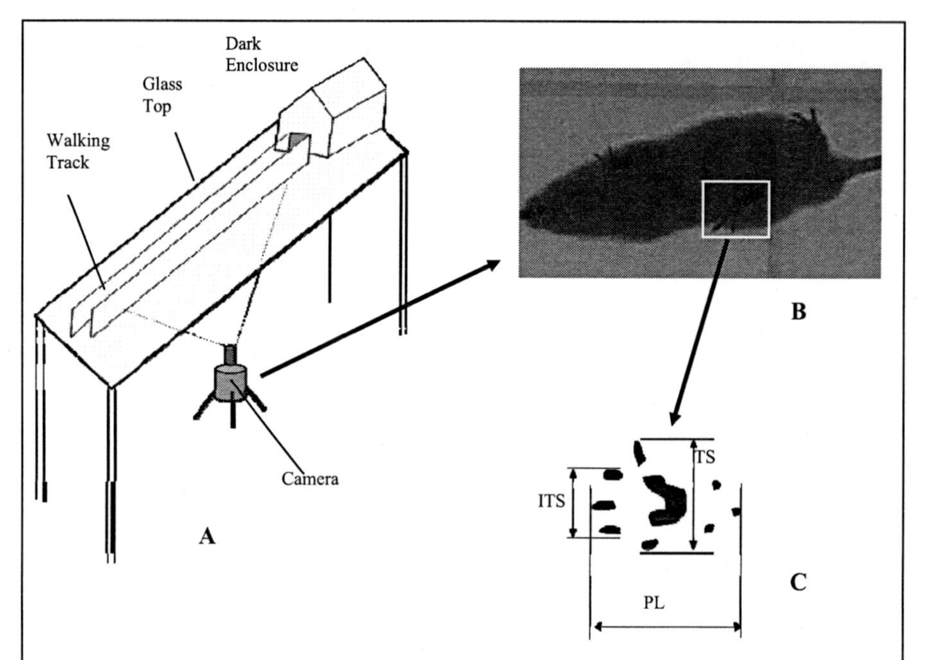

Figure 3-4: Schematic showing the measurement of rodent sciatic function. **A.** Schematic of imaging setup. **B.** An example of a captured footprint during gait. **C.** The three variables measured from the footprint. ITS= Intermediate Toe Spread, TS= Toe Spread, and PL= Print Length.

$$SFI = -38.3 \times PLF + 109.5 \times TSF + 13.3 \times ITF - 8.8$$

A SFI value of 0 is considered normal. A SFI of -100 indicates total impairment, as it would be for a complete transection of the sciatic nerve. If there is functional recovery due to regeneration techniques, the SFI should increase from -100 towards 0. So SFI can be used as a parameter for measuring functional recovery post-injury.

3.2.3.6. Statistical analysis

The data were expressed as mean ± SEM. A one-way ANOVA was used to compare the data between groups for histological analysis and functional analysis. A p-value < 0.05 was considered statistically significant.

3.3. Results

3.3.1. LN-1 and NGF distribution in Isotropic and Anisotropic Scaffolds

LN-1 Concentration in Isotropic Scaffolds

Bradford assay (BIO-RAD, Hercules, CA) was used to determine the amount of LN-1 coupled to agarose gel. Efficiency of the conjugation technique was about 10-15%. LN-1-agarose gel with LN-1 concentration of 66 μg/ml of 1% agarose solution was used for *in vivo* experiments.

NGF Release Study

In vitro NGF-release from LMTs was followed up to 18 days. The amount of NGF released into the supernatant (PBS solution) was determined by NGF ELISA kit (Figure 3-5). For the first 6 days, the cumulative amount of NGF released was linear and rapid, and for the next 12 days the cumulative release still increased linearly, but less rapidly. The amount of NGF released at 18th day was still much above the detection limit for the ELISA kit, indicating that NGF could be released for more than 18 days from the LMTs.

LN-1 distribution in anisotropic scaffolds

LN-1 distribution in continuous-gradient anisotropic scaffolds was determined by LN-1 ELISA. The 20-mm scaffolds were cut transversely into 4 parts, 5-mm each, and LN-1 ELISA was performed. In Figure 3-6, LN-1 concentration distribution is plotted along the length of NGC. The average LN-1 concentration in a 5-mm section has been plotted at mid-length (i.e. 2.5-mm, 7.5-mm, etc.). LN-1 distribution closely followed a second-order polynomial. The continuous-gradient anisotropic scaffolds with LN-1 distribution close to that of isotropic scaffolds (4 layers of gel with 33 μg/ml, 25 μg/ml, 16.5 μg/ml, and 8.25 μg/ml, respectively, Figure 3-3A) were used for *in vivo* studies.

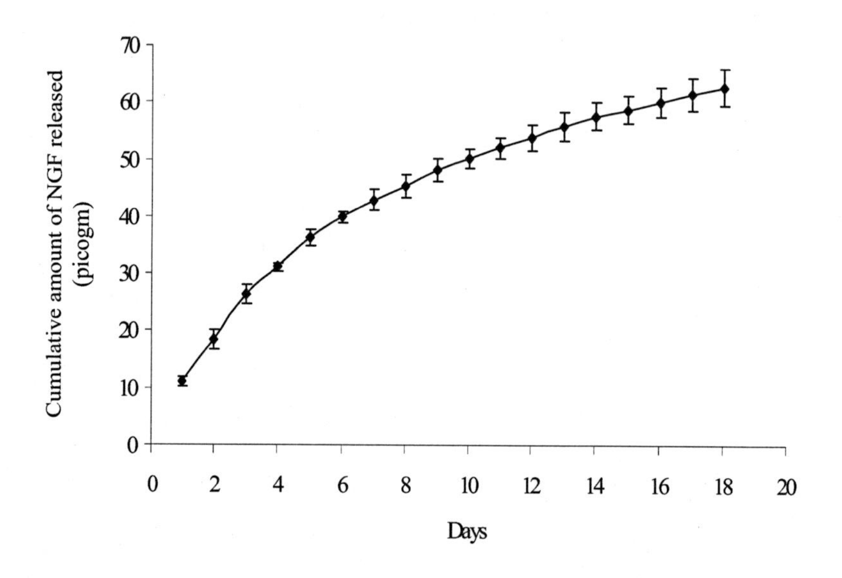

Figure 3-5: NGF-release study, in vitro. NGF-release from LMTs was followed up to 18 days. For the first 6 days, cumulative amount of NGF-released was linear and rapid. For the next 12 days, the release was less rapid but still increased linearly with time (n = 2).

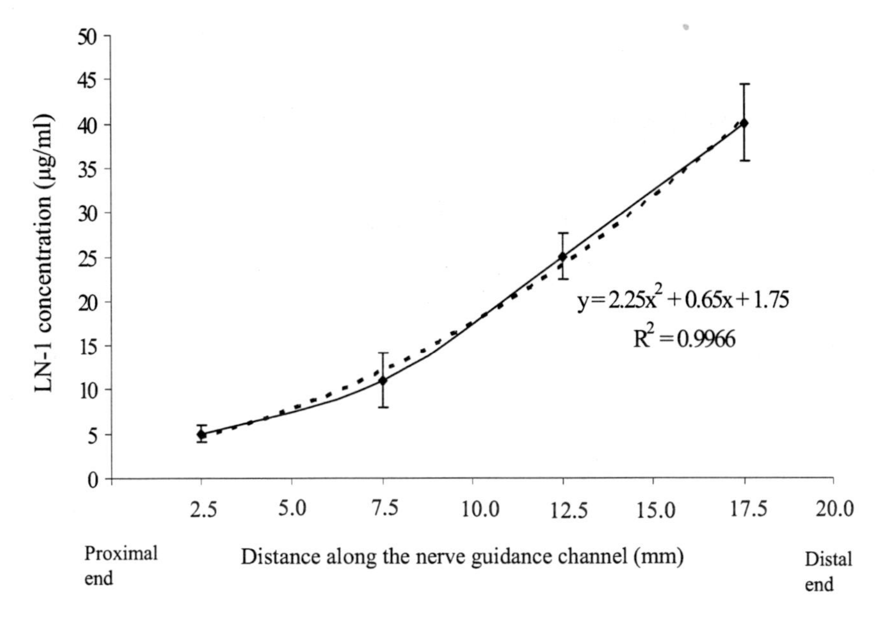

Figure 3-6: LN-1 concentration distribution in continuous-gradient anisotropic scaffolds. 20-mm long nerve guidance channels with continuous-gradients of LN-1 were cut into 4 parts, each 5-mm long, and their LN-1 content was determined. LN-1 distribution followed a second-order polynomial (n = 3).

3.3.2: Histological Analysis for Nerve Regeneration

3.3.2.1. Presence of Regenerating Axons in Nerve Implants

Presence of myelinated axons was indentified by OsO_4 and toulidine blue staining. Nerve implants with positive staining for myelinated axons were counted as successful in promoting axonal regeneration. A total of 10 groups were studied and only 3 groups had axonal regeneration through the implants, as determined by histological analysis (Table 3-2, below). The 3 groups were, rats with *nerve grafts*, *LN(SG)+NGF(SG)* (step-gradient of LN-1 and NGF), and *LN(CG)+NGF(SG)* (continuous-gradient of LN-1 and step-gradient of NGF). The success rate was highest for nerve grafts (83.3%), followed by LN(SG)+NGF(SG) (44.4%) and LN(CG)+NGF(SG) (37.5%) (Table 3-2). Rest of the histological analysis (axonal area, number of myelinated axons, etc.) could be done for these 3 groups only.

Table 3-2: Summary of Nerve Regeneration in Control and Experimental Groups

Groups	Total # of rat	# of rats with axon regeneration	% regeneration
Saline	5	0	0
Plain agarose	5	0	0
LN(U)	5	0	0
LN(U)+NGF(U)	10	0	0
LN(U)+NGF(SG)	9	0	0
LN(SG)+NGF(U)	11	0	0
LN(SG)+NGF(SG)	**9**	**4**	**44.4**
LN(CG)+NGF(U)	7	0	0
LN(CG)+NGF(SG)	**8**	**3**	**37.5**
Nerve graft	**6**	**5**	**83.3**

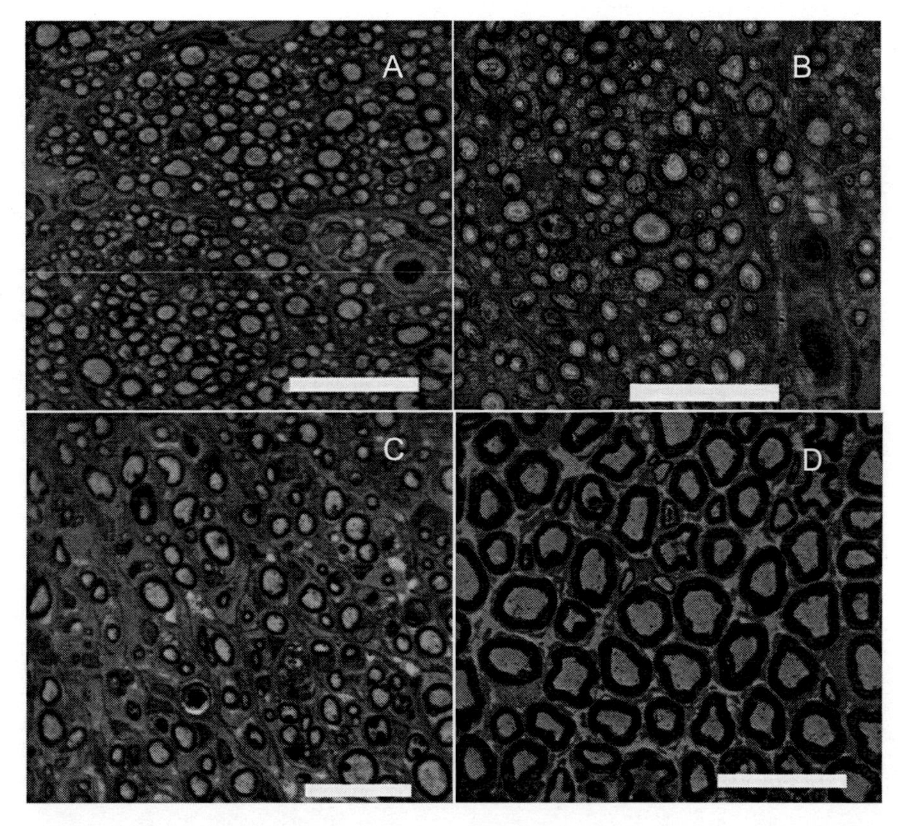

Figure 3-7: Light micrographs of nerve sections of groups with successful regeneration. Toluidine-blue stained cross-sections at mid-length (10-mm) of nerve implants are shown here. Anisotropic scaffolds with step-gradient (A) and continuous-gradient (B) of LN-1, have higher density of axons and smaller-diameter axons than nerve grafts (C) and normal nerve (D). Scale bar = 25μm.

The photographs of histology specimen of cross sections at the mid-point of nerve guidance channels/autografts/normal nerve are shown in Figure 3-7 (A-D). The gross structure of the regenerated nerves was similar to the normal sciatic nerve. Regenerated axons were packed in bundles to form fascicles, and several fascicles were enclosed in the connective tissues of epineurium. Large numbers of nuclei from Schwann cells clearly surrounded the regenerated axons. There were abundant blood vessels in the epineurium of regenerated nerves. The implanted agarose gels were completely resorbed without any trace.

3.3.2.2. Axonal Area of Regeneration:

The area occupied by regenerating axons was outlined and measured using ImagePro® software. The area of regeneration was determined at mid-length (10-mm) and distal end (17-mm) of the implants for the 3 groups which showed presence of myelinated axons. The area was significantly higher for the nerve grafts than the anisotropic scaffolds. However, there was no significant difference between the two anisotropic scaffolds, LN(SG)+NGF(SG) and LN(CG)+NGF(SG) group (Figure 3-8). For all the 3 groups, there was no significant difference between the area of regeneration at mid-length (10-mm) and at distal end (17-mm).

3.3.2.3. Total Number of Myelinated Axons and the Density of Myelinated Axons

For all 3 groups, there was no significant difference between the total number of myelinated axons at mid-length (10-mm) and at distal end (17-mm) (Figure 3-9). At mid-length, the total number of myelinated axons in the nerve grafts and the

LN(SG)+NGF(SG) (step-gradient) implants were not significantly different. However, at mid-length, the number of myelinated axons in the LN(CG)+NGF(SG) (continuous-gradient) implants was less than that of nerve grafts. At distal end (17-mm), there was no significant difference between the 3 groups. The number of myelinated axons in a normal sciatic nerve has also been shown here. The total number of myelinated axons in the nerve grafts and the LN(SG)+NGF(SG) group is comparable to a normal sciatic nerve (Figure 3-9). However, the LN(CG)+NGF(SG) group has lower number of myelinated axons than a normal nerve.

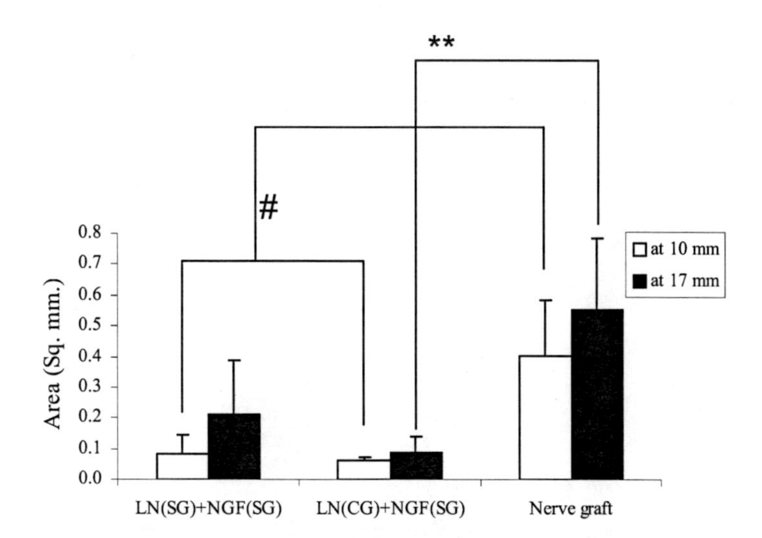

Figure 3-8: Area of nerve regeneration. For all three groups, there was no significant difference between area of regeneration at distal end and at the middle. At midpoint, nerve graft has higher area of regeneration than either anisotropic scaffold group (# p-value < 0.05). However, at distal end, area in nerve graft is higher compared to continuous-gradient group only (** p-value<0.05).

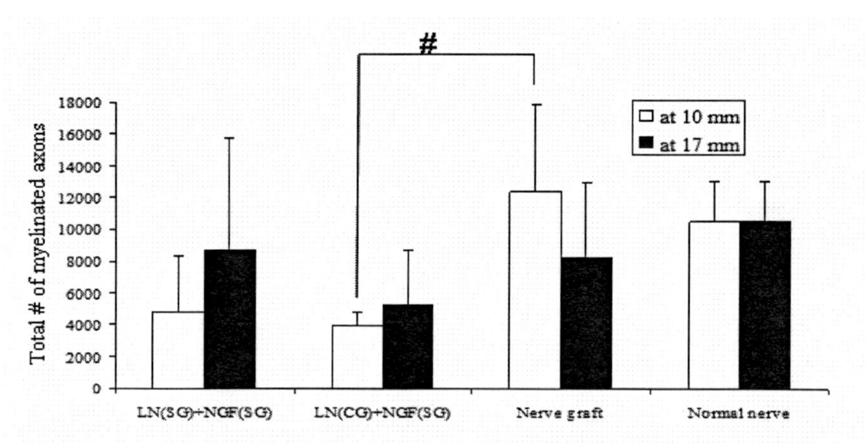

Figure 3-9: Total number of myelinated axons. There was no significant difference between step-gradients and nerve grafts. There was no significant difference between the two gradient scaffolds themselves. At mid-length, the continuous-gradient scaffolds had lower number of myelinated axons than nerve grafts (# p-value < 0.05).

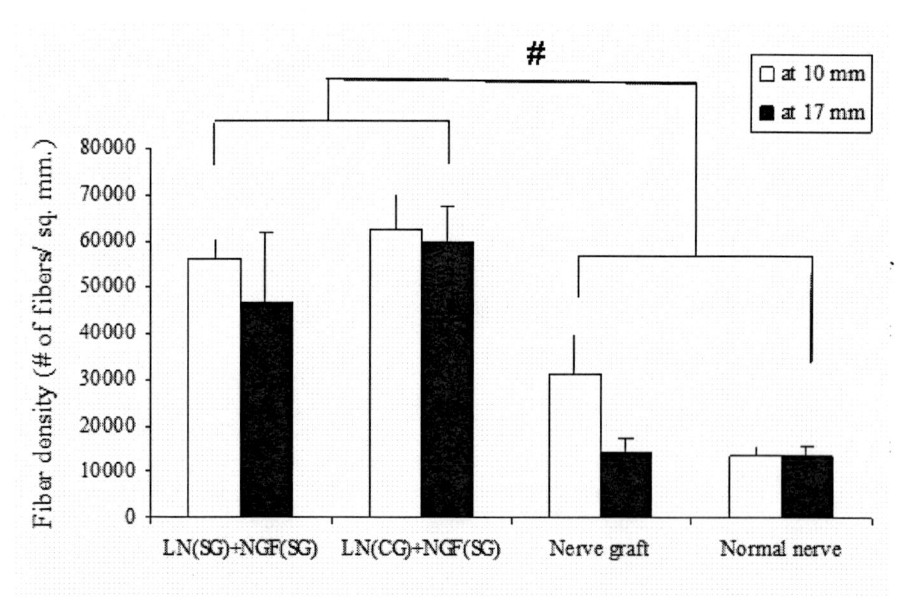

Figure 3-10: Density of myelinated axons. The regenerated nerve fibers are more densely packed in anisotropic scaffolds than nerve grafts and normal nerve (# p-value<0.05).

The regenerated nerve cables were more densely packed in the anisotropic scaffolds than the nerve grafts and a normal nerve (Figure 3-10). The fiber density in anisotropic scaffolds is almost twice that of nerve grafts and five times that of a normal nerve.

3.3.2.4. Nerve Fiber Diameter Distribution

Nerve fiber diameter was measured using ImagePro® software to determine the diameter distribution of regenerated nerve fibers, at mid-length and distal end. The diameter distribution followed a bell-shaped curve (Figure 3-11a). The peak of the curve, i.e., the maximum number of nerve fibers, was in the 1-2 μm diameter range for the nerve grafts as well as the anisotropic scaffolds. However, the fiber-diameter distribution in a normal nerve was a bell-shaped curve with a much wider spread. In a normal nerve the maximum number of fibers had a diameter range from 2-5 μm. The fiber diameter distribution pattern at distal end (17-mm) was similar to that at mid-length (10-mm) (Figure 3-11b). Hence, the nerve implants have more regenerated nerve fibers in the smaller diameter range than a normal nerve, but fewer nerve fibers in the larger diameter range than a normal nerve.

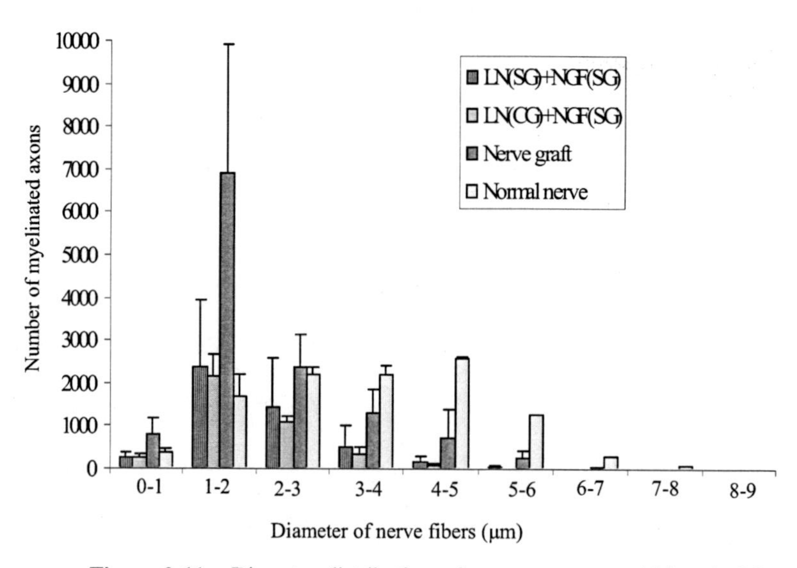

Figure 3-11a: Diameter distribution of nerve axons at mid-length (10-mm). For each group the diameter distribution is a bell-shaped curve. For all the nerve implants, the maximum number of axons have diameter of 1-2 µm. However, in a normal nerve the maximum diameter distribution is more wide, from 1-5 µm.

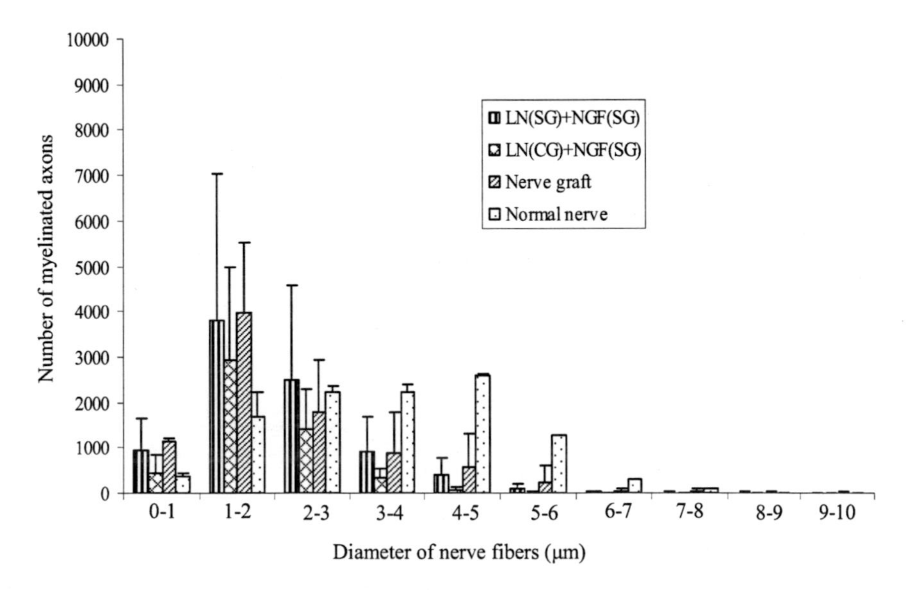

Figure 3-11b: Diameter distribution of nerve axons at distal end (17-mm). For each group the diameter distribution is a bell-shaped curve. For all the nerve implants, the maximum number of axons have diameter of 1-2 μm. However, in a normal nerve the maximum diameter distribution is wider, from 1-5 μm.

3.3.3. Relative Gastrocnemius Muscle Weight

The gastrocnemius muscle starts to atrophy after sciatic nerve injury, and hence, its mass is directly proportional to the extent of sciatic nerve re-innervation. There was no significant difference in RGMW values between LN(SG)+NGF(SG) (step-gradients) group and nerve grafts (Figure 3-12). Rest of the groups had significantly lower RGMW values. Although, LN(CG)+NGF(SG) (continuous-gradient) group showed RGMW values close to that of nerve grafts, it was still significantly lower.

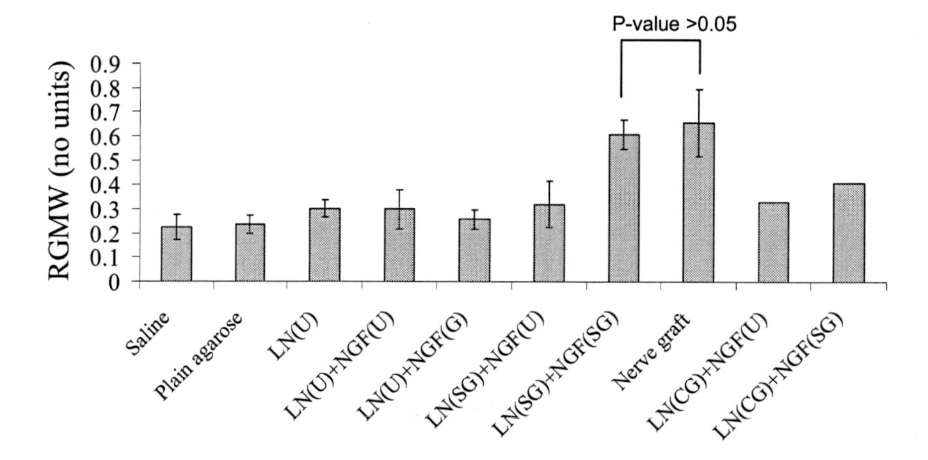

Figure 3-12: Relative gastrocnemius muscle weight. Implants with step-gradients and nerve grafts had comparable RGMW (p-value > 0.05). The rest of the implant groups had significantly lower RGMW.

3.3.4. Neuromuscular junction, synaptic vesicles and neurofilament staining

25 μm thick sections of gastrocnemius muscle from normal (uninjured) rats, and treatment groups (nerve graft and anisotropic scaffolds) were stained for acetylcholine receptors (neuromuscular junctions, NMJ), neurofilament and synaptic vesiscles. NMJ (red, Figure 3-13(A)) which were positive for neurofilament staining (green, Figure 3-13(B)) and/or synaptic vesicles (green, Figure 3-13(C)) were counted and compared for the various groups. In a normal rat, 38.5% of NMJs were positive for neurofilament and/or synaptic vesiscles. However, in the treated groups, there were no NMJs positive for neurofilament and/or synaptic vesicle staining, indicating that the regenerated axons had not formed neuromuscular junctions.

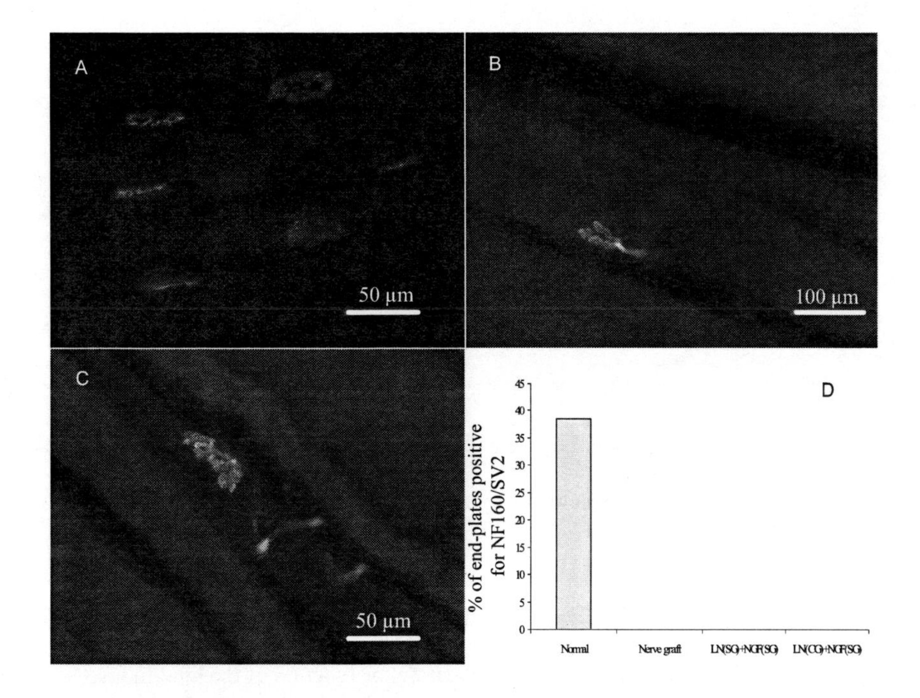

Figure 3-13: Staining for acetylcholine receptors, synaptic vesicles and neurofilaments. In a normal uninjured rat, nicotinic acetylcholine receptors (red, A), neurofilaments (green, B) and synaptic vesicles (c) could be identified. In normal rat, 38.5% of end-plates were positive for NF160 and/or SV2 staining (D). However, treated rats, nerve grafts and anisotropic scaffolds, did not have any end-plates positive for NF160 or SV2 staining.

3.3.5. Electrophysiological recordings

4-months post-implantation, to evaluate the regeneration and maturation of the nerve in the grafts, CAPs were recorded between the proximal and distal ends of the grafts. The distance between the stimulating and recording electrode was 60-70 mm. Action potentials were also recorded from normal sciatic nerves. Action potentials could be recorded in nerve grafts and normal nerves (Figure 3-14). However, action potentials could not be recorded from anisotropic scaffolds, even after raising the stimulating voltage 20-times compared to the normal nerves. Similarly, action potential could not be evoked in other experimental and control groups.

3.3.6. Walking Track Analysis for SFI measurement

SFI is an indication of recovery of motor function of the hind limbs. SFI value of -100 indicates total impairment and value of 0 indicates normal condition. The SFI values of all the eight groups were close to -100, indicating that the nerve regeneration observed in histological analysis did not result in significant improvement in motor function (Figure 3-15).

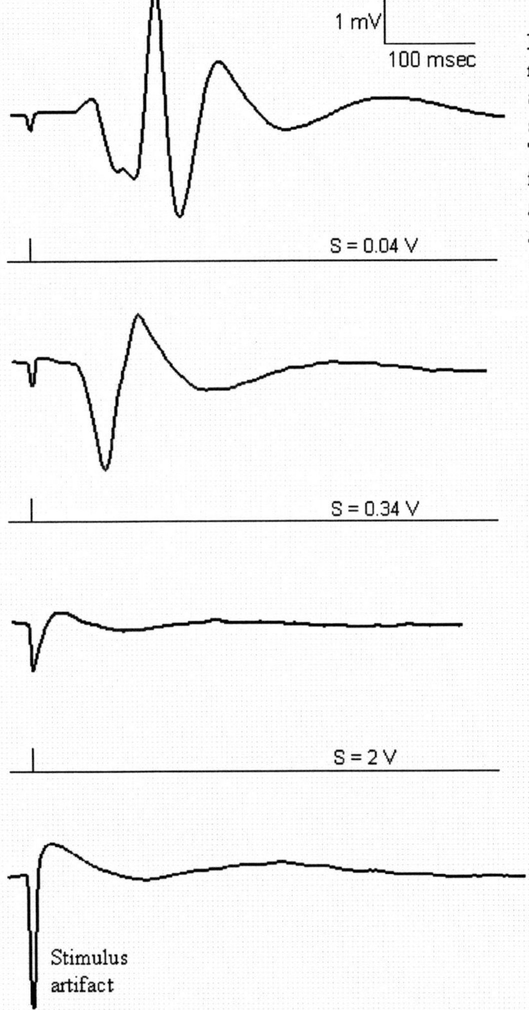

Figure 3-14: Electrophysiological recordings. For normal nerve (A) and nerve grafts (B), S represents stimulus voltage 1.5 x threshold. The initial stimulus artifact is followed by action potential. Step-gradient (C) and continuous-gradient (D) implants were stimulated up to 2V, but failed to evoke action potential.

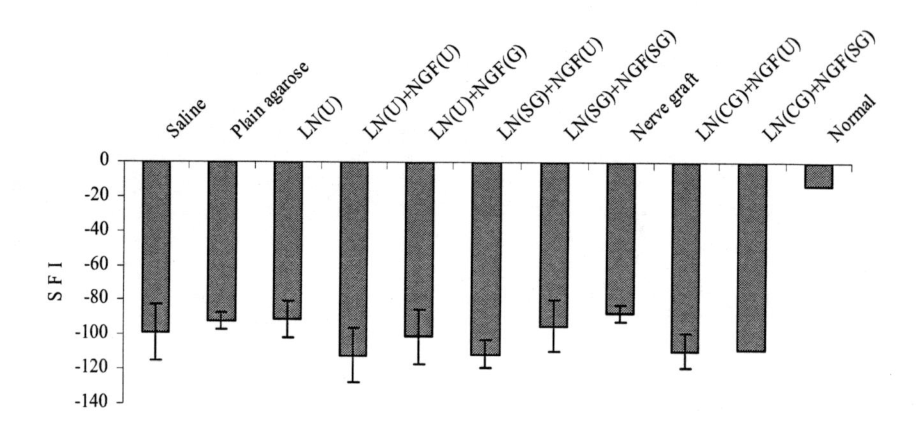

Figure 3-15: Sciatic functional index. The values of SFI for almost all groups were close to -100. There was no significant difference between experimental and control groups. "Normal" indicates an uninjured rat.

3.4. Discussion

During embryonic development of nervous system, the developing axons use a variety of haptotactic and chemotactic cues to find their target organs [135]. Some of these cues are presented in a gradient fashion in order to direct the growing axons towards their target tissues [16,17]. ECM proteins, such as laminin and fibronectin, and neurotrophic factors, such as NGF and BDNF, have been used in several studies to promote nerve regeneration *in vivo* [4,7,26]. However, these neuro-stimulatory molecules have not been presented in a gradient fashion as observed *in vivo*. To simulate *in vivo* conditions for axonal guidance and regeneration, we present a novel technique to synthesize gradients of ECM proteins and neurotrophic factors for nerve regeneration *in vivo*. Here we show that anisotropic agarose gel scaffolds with gradients of LN-1 and

NGF promote enhanced axonal regeneration, as compared to isotropic scaffolds with uniform concentrations of LN-1 and NGF, across a challenging 20-mm nerve gap in rats. In the nerve injury model studied here, regeneration is limited and does not occur unless the nerve guidance channels have gradients of LN-1 and NGF (step- or continuous-). Moreover, gradients of both LN-1 and NGF are needed to promote neurite extension. Anisotropic scaffolds with gradient of only LN-1 or only NGF, with the other component at uniform concentration, did not promote axonal regeneration across the challenging 20-mm nerve gap. The enhanced axonal regeneration with anisotropic scaffolds, however, did not result in enhanced motor function recovery as measured by walking-track analysis, relative gastrocnemius muscle weight and electrophysiology.

LN-1 is an appropriate candidate as the protein of choice for PNS regeneration since it has been shown to be a potent stimulator of neuronal cell migration and neurite outgrowth [13,124,126]. In most neuronal tissues, development of the axons and formation of new synapses is preceded by migration of the neuronal cell bodies to the appropriate regions of the brain. It has been proposed that LN facilitates these processes in several neural systems [19,98,167]. The spatio-temporally controlled expression of LN-1 in the developing peripheral nervous system, as well as the visual pathway and the cerebellum, suggests that LN plays a role in axon outgrowth and guidance [99,111,125]. LN expression has been found to be essential for ventral turning of axons during development. Blocking nidogen-binding motifs on LN prevents turning of axons although it does not affect growth [17], indicating that LN might play a role in orientation of axons in some systems. NGF has been shown to stimulate and promote the survival of sensory ganglia and nerves, including spinal sensory nerves and sciatic nerves [95,145].

Griffin et al. [59] have shown concentration-dependent neurite extension/retraction with NGF levels. NGF has been shown to prevent the death of axotomized sensory neurons completely following exogenous administration [123]. There is enough evidence to show that gradients of neurotrophins help in guiding developing as well as regenerating axons in a wide variety of neuronal systems [63,64]. Although many *in vitro* studies have shown influence of gradients of LN-1 and NGF for neurite extension, to our knowledge, there are no *in vivo* studies with gradients, due to the difficulties in making and sustaining gradients *in vivo*.

In our study, we present a diffusion technique to first make gradients of LN-1 in agarose gel and then immobilize it using photochemical cross-linking. Immobilization of LN-1 imitates *in vivo* conditions where LN-1 is tethered to the ECM. Gradients of NGF were designed using slow-release LMTs, which imitate *in vivo* conditions. Two kinds of LN-1 gradients were designed, a step-gradient and the other a continuous-gradient, to promote nerve regeneration. However, the NGF gradient was always a continuous-gradient due to the diffusion of NGF. The step- and continuous-gradient scaffolds of LN-1 were designed such that the average LN-1 concentration in both conditions is very close. However, the actual concentration of LN-1 that the regenerating axons will sense dynamically in each condition will be different because of the difference in gradient profile. For example, over a 0-5 mm distance in step-gradients, axons will encounter a uniform concentration of LN-1 (8.25 μg/ml), whereas in a continuous-gradient, an increasing concentration profile of LN-1 will be encountered, but with an average LN-1 concentration of 5.1 μg/ml. The LN-1 concentration range selected for making LN-1 step-gradients was based on the maximum concentration of LN-1 that could be immobilized. However, this concentration range may not be the optimal range. A more thorough study will be needed to determine the optimal LN-1 concentration range for nerve regeneration.

Nerve regeneration over a 20-mm nerve gap in rats is challenging, and could be the reason for the drastic results observed in nerve histology. Nerve regeneration was observed only when the scaffolds had gradients of both LN-1 and NGF. Scaffolds with gradient of only LN-1 or only NGF, with the other component at uniform concentration, were not able to promote nerve regeneration. This suggests a synergistic effect between LN-1 and NGF, as suggested in a previous study [163].

In relative gastrocnemius muscle weight study, the anisotropic scaffolds with step-gradients of LN-1 were able to match the autografts for RGMW. Rest of the groups had significantly lower RGMW. This again suggests that gradients of both LN-1 and NGF are needed to promote nerve regeneration across a challenging nerve gap of 20-mm. Lower RGMW results with the continuous-gradient scaffolds, as compared to the step-gradient scaffolds, could be due to non-optimal concentration range of LN-1 in the continuous-gradient scaffolds. Even though both groups, nerve grafts and step-gradient scaffolds, show significant improvement in RGMW, they were still lower than the native, well-innervated muscle. It is possible that the regenerating axons take longer than 4-months to traverse the 20-mm nerve gap and the distal nerve stump, to form new neuromuscular junctions and cause muscle regeneration. A study with longer duration of regeneration may produce higher values of RGMW for both groups.

To determine formation of new neuromuscular junctions (NMJs), gastrocnemius muscle sections were stained for acetylcholine receptors positive for neurofilament and/or synaptic vesicles. Although the nerve grafts and anisotropic scaffolds were positive for some neurofilament and acetylcholine receptor staining, the regenerating axons did not connect to the acetylcholine receptors. This indicates that the regenerating axons had not traveled far enough to form new NMJs. A longer observation period (more than 4-months) might have lead to formation of new NMJs.

Electrophysiology studies were done to determine the maturity and extent of nerve regeneration through the nerve grafts. Normal sciatic nerve (uninjured) requires only 0.02V stimulation to evoke action potential. However, the nerve grafts requires 0.22V stimulation to evoke action potential. This could be explained based on the fiber diameter distribution observed in histological analysis. For nerve grafts and anisotropic scaffolds, the maximum number of regenerating axons were in the 1-2 μm range, whereas, for a normal nerve the range was 1-5 μm. Since the normal nerves have more nerve fibers with larger diameters, they are easy to stimulate than nerve grafts. Although regenerating nerve fibers were observed by histological analysis in anisotropic scaffolds, action potentials could not be evoked. This suggests two possibilities: either the regenerating nerve fibers are very small and hence could not be stimulated, or the regenerating nerve fibers did not reach the site of recording electrode. The recording electrode was 60-70 mm distal to the stimulating electrode, and hence 40-50 mm from the distal end of nerve graft (since grafts are 20 mm long). It is possible that in case of anisotropic scaffolds, the regenerating axons were not able to traverse the 40-50 mm length up to the recording electrode yet, and hence no action potentials were recorded. This suggests that the rate of nerve regeneration is slower in anisotropic scaffolds as compared to nerve grafts. However, since we did not do histology of the nerve at the recording site, it is not possible to say if the regenerating axons failed to reach the recording site.

SFI data shows that there is no significance difference between the various groups, with all the groups having SFI values close to -100, suggesting that there was no motor functional recovery as measured by SFI. It is possible that 4-months are not enough for sufficient number of the regenerating axons to innervate the gastrocnemius muscle and improve motor function. A study with longer observation period might have resulted in better functional recovery. Also, more sensitive quantitative methods including sciatic functional index, gait-stance duration, ankle-kinematics and toe out angle (TAO) could be used. A combination of tests, each examining particular components of recovered sensorimotor function may lead to better overall assessment of rat sciatic nerve regeneration [154].

In summary, here we present a novel technique to synthesize anisotropic scaffolds with gradients of ECM proteins and neurotrophic factors for *in vivo* nerve regeneration. These scaffolds closely mimic *in vivo* conditions for presenting these molecules. The results also demonstrate that nerve regeneration across a challenging 20-mm nerve gap is possible only when gradients of both LN-1 and NGF are present. Although, the anisotropic scaffolds showed nerve regeneration comparable to nerve grafts, the nerve grafts resulted in better functional outcome as measured by RGMW and electrophysiology. Optimizing the concentration of LN-1 and NGF might further improve the performance of anisotropic scaffolds.

3.5. References

[1] Adams, D.N., Kao, E.Y., Hypolite, C.L., Distefano, M.D., Hu, W.S. and Letourneau, P.C., Growth cones turn and migrate up an immobilized gradient of the laminin IKVAV peptide, *J Neurobiol*, 62 (2005) 134-47.

[2] Aebischer, P., Guenard, V. and Brace, S., Peripheral nerve regeneration through blind-ended semipermeable guidance channels:effect of molecular weight cutoff, *J Neurosci*, 9 (1989a) 3590-3595.

[3] Aebischer, P., Guenard, V. and Valentini, R.F., The morphology of regenerating peripheral nerves is modulated by the surface microgeometry of polymeric guidance channels, *Brain Res*, 531 (1990) 211-8.

[4] Aebischer, P., Salessiotis, A.N. and Winn, S.R., Basic fibroblast growth factor released from synthetic guidance channels facilitates peripheral nerve regeneration across long nerve gaps, *J Neurosci Res*, 23 (1989) 282-9.

[5] Aebischer, P., Valentini, R.F., Dario, P., Domenici, C. and Galletti, P.M., Piezoelectric guidance channels enhance regeneration in the mouse sciatic nerve after axotomy, *Brain Res*, 436 (1987) 165-8.

[6] Aebischer, P., Valentini, R.F., Winn, S.R. and Galletti, P.M., The use of a semi-permeable tube as a guidance channel for a transected rabbit optic nerve, *Prog Brain Res*, 78 (1988) 599-603.

[7] Ahmed, Z., Brown, R.A., Ljungberg, C., Wiberg, M. and Terenghi, G., Nerve growth factor enhances peripheral nerve regeneration in non-human primates, *Scand J Plast Reconstr Surg Hand Surg*, 33 (1999) 393-401.

[8] Albert, E., Einige Operationen an Nerven, *Wien Med*, 26 (1885) 1285.

[9] Amoh, Y., Li, L., Campillo, R., Kawahara, K., Katsuoka, K., Penman, S. and Hoffman, R.M., Implanted hair follicle stem cells form Schwann cells that support repair of severed peripheral nerves, *Proc Natl Acad Sci U S A*, 102 (2005) 17734-8.

[10] Ansselin, A.D., Fink, T. and Davey, D.F., Peripheral nerve regeneration through nerve guides seeded with adult Schwann cells, *Neuropathol Appl Neurobiol*, 23 (1997) 387-98.

[11] Baier, H. and Bonhoeffer, F., Axon guidance by gradients of a target-derived component, *Science*, 255 (1992) 472-5.

[12] Barde, Y.A., Trophic factors and neuronal survival, *Neuron*, 2 (1989) 1525-34.

[13] Baron-van Evercooren, A., Kleinman, H.D., Ohno, S., Marangos, P., Schwartz, J.P. and Dubois-Dalcq, M.E., Nerv growth factor, laminin and fibronectin promote nerve growth in human fetal sensory ganglia cultures, *J Neurosci. Res.* (1982) 179-183.

[14] Battiston, B., Tos, P., Cushway, T.R. and Geuna, S., Nerve repair by means of vein filled with muscle grafts I. Clinical results, *Microsurgery*, 20 (2000) 32-36.

[15] Bellamkonda, R.V., Ranieri, J.P., Bouche, N. and Aebischer, P., Hydrogel-based three-dimensional matrix for neural cells, *J Biomed Mater Res*, 29 (1995) 663-71.

[16] Bonhoeffer, F. and Huf, J., In vitro experiments on axon guidance demonstrating an anterior-posterior gradient on the tectum, *Embo J*, 1 (1982) 427-31.

[17] Bonner, J. and O'Connor, T.P., The permissive cue laminin is essential for growth cone turning in vivo, *J Neurosci*, 21 (2001) 9782-91.

[18] Brandt, J., Dahlin, L.B. and Lundborg, G., Autologous tendons used as grafts for bridging peripheral nerve defects, *J Hand Surg [Br]*, 24 (1999) 284-90.

[19] Bronner-Fraser, M., Stern, C.D. and Fraser, S., Analysis of neural crest cell lineage and migration, *J Craniofac Genet Dev Biol*, 11 (1991) 214-22.

[20] Bryan, D.J., Wang, K.-K. and Chakalis-Haley, D.P., Effect of Schwann cells in the enhancement of peripheral nerve regeneration, *J. Reconstr. Microsurg.*, 12 (1996) 439-446.

[21] Bunge, R.P., Expanding roles for the Schwann cell: ensheathment, myelination, trophism and regeneration, *Curr Opin Neurobiol*, 3 (1993) 805-9.

[22] Cajal, S.R.y., Degeneration and regeneration of the nervous system, *Hafner* (1928).

[23] Cao, X. and Shoichet, M.S., Defining the concentration gradient of nerve growth factor for guided neurite outgrowth, *Neuroscience*, 103 (2001) 831-40.

[24] Cao, X. and Shoichet, M.S., Investigating the synergistic effect of combined neurotrophic factor concentration gradients to guide axonal growth, *Neuroscience*, 122 (2003) 381-9.

[25] Champion, S., Imhof, B.A., Savagner, P. and Thiery, J.P., The embryonic thymus produces chemotactic peptides involved in the homing of hemopoietic precursors, *Cell*, 44 (1986) 781-90.

[26] Chen, Y.S., Hsieh, C.L., Tsai, C.C., Chen, T.H., Cheng, W.C., Hu, C.L. and Yao, C.H., Peripheral nerve regeneration using silicone rubber chambers filled with collagen, laminin and fibronectin, *Biomaterials*, 21 (2000) 1541-7.

[27] Chiu, D.T., Janecka, I., Krizek, T.J., Wolff, M. and Lovelace, R.E., Autogenous vein graft as a conduit for nerve regeneration, *Surgery*, 91 (1982) 226-33.

[28] Chiu, D.T., Lovelace, R.E., Yu, L.T., Wolff, M., Stengel, S., Middleton, L., Janecka, I.P., and Krizek, T.J., Comparative electrophysiological evaluation of nerve grafts and autogenous vein grafts as nerve conduits: an experimental study, *J. Reconstr. Microsurg.*, 4 (1988) 303-309.

[29] Chiu, D.T. and Strauch, B., A prospective clinical evaluation of autogenous vein grafts used as a nerve conduit for distal sensory nerve defects of 3 cm or less, *Plast Reconstr Surg*, 86 (1990) 928-34.

[30] Choi, B.H., Zhu, S.J., Kim, B.Y., Huh, J.Y., Lee, S.H. and Jung, J.H., Transplantation of cultured bone marrow stromal cells to improve peripheral nerve regeneration, *Int J Oral Maxillofac Surg*, 34 (2005) 537-42.

[31] Choi, B.H., Zhu, S.J., Kim, S.H., Kim, B.Y., Huh, J.H., Lee, S.H. and Jung, J.H., Nerve repair using a vein graft filled with collagen gel, *J Reconstr Microsurg*, 21 (2005) 267-72.

[32] Clark, P., Britland, S. and Connolly, P., Growth cone guidance and neuron morphology on micropatterned laminin surfaces, *J Cell Sci*, 105 (Pt 1) (1993) 203-12.

[33] Condic, M.L. and Letourneau, P.C., Ligand-induced changes in integrin expression regulate neuronal adhesion and neurite outgrowth, *Nature*, 389 (1997) 852-856.

[34] Cornbrooks, C.J., Carey, D.J., McDonald, J.A., Timpl, R. and Bunge, R.P., In vivo and in vitro observations on laminin production by Schwann cells, *Proc Natl Acad Sci U S A*, 80 (1983) 3850-4.

[35] Dahlin, L.B. and Lundborg, G., Use of tubes in peripheral nerve repair, *Neurosurg Clin N Am*, 12 (2001) 341-52.

[36] Danielsen, N., Dahlin, L.B., Lee, Y.F. and Lundborg, G., Axonal growth in mesothelial chambers. The role of the distal nerve segment, *Scand J Plast Reconstr Surg*, 17 (1983) 119-25.

[37] de Medinaceli, L., Freed, W.J. and Wyatt, R.J., An index of the functional condition of rat sciatic nerve based on measurements made from walking tracks, *Exp Neurol*, 77 (1982) 634-43.

[38] Derby, A., Engleman, V.W., Frierdich, G.E., Neises, G., Rapp, S.R. and Roufa, D.G., Nerve growth factor facilitates regeneration across nerve gaps: morphological and behavioral studies in rat sciatic nerve, *Exp Neurol*, 119 (1993) 176-91.

[39] Dertinger, S.K., Jiang, X., Li, Z., Murthy, V.N. and Whitesides, G.M., Gradients of substrate-bound laminin orient axonal specification of neurons, *Proc Natl Acad Sci U S A*, 99 (2002) 12542-7.

[40] Dillon, G.P., Yu, X. and Bellamkonda, R.V., The polarity and magnitude of ambient charge influences three-dimensional neurite extension from DRGs, *J Biomed Mater Res*, 51 (2000) 510-9.

[41] Dillon, G.P., Yu, X., Sridharan, A., Ranieri, J.P. and Bellamkonda, R.V., The influence of physical structure and charge on neurite extension in a 3D hydrogel scaffold, *J Biomater Sci Polym Ed*, 9 (1998) 1049-69.

[42] Dodla, M.C. and Bellamkonda, R.V., Anisotropic scaffolds facilitate enhanced neurite extension in vitro, *J Biomed Mater Res A*, 78 (2006) 213-21.

[43] Dubey, N., Letourneau, P.C. and Tranquillo, R.T., Neuronal contact guidance in magnetically aligned fibrin gels: effect of variation in gel mechano-structural properties, *Biomaterials*, 22 (2001) 1065-75.

[44] Edelman, E.R., Mathiowitz, E., Langer, R. and Klagsbrun, M., Controlled and modulated release of basic fibroblast growth factor, *Biomaterials*, 12 (1991) 619-26.

[45] Edgar, D., Nerve growth factors and molecules of the extracellular matrix in neuronal development, *J Cell Sci Suppl*, 3 (1985) 107-13.

[46] English, A.W., Meador, W. and Carrasco, D.I., Neurotrophin-4/5 is required for the early growth of regenerating axons in peripheral nerves., *Eur. J. Neurosci.*, 21 (2005) 2624-34.

[47] Esch, T., Lemmon, V. and Banker, G., Local presentation of substrate molecules directs axon specification by cultured hippocampal neurons, *J Neurosci*, 19 (1999) 6417-6426.

[48] Evans, G.R., Peripheral nerve injury: a review and approach to tissue engineered constructs, *Anat Rec*, 263 (2001) 396-404.

[49] Fawcett, J.W.a.K., R.J., Peripheral nerve regeneration, *Ann. Rev. Neurosci.*, 13 (1990) 43-60.

[50] Ferrari, F., De Castro Rodrigues, A., Malvezzi, C.K., Dal Pai Silava, M. and Padvoni, C.R., Inside-out vs. standard vein graft to repair a sensory nerve in rats, *Anat Rec*, 256 (1999) 227-232.

[51] Fu, S.Y. and Gordon, T., The cellular and molecular basis of peripheral nerve regeneration, *Mol Neurobiol*, 14 (1997) 67-116.

[52] Gallo, G., Lefcort, F.B. and Letourneau, P.C., The trkA receptor mediates growth cone turning toward a localized source of nerve growth factor, *J Neurosci*, 17 (1997) 5445-54.

[53] Garcia-Alonso, L., Fetter, R.D. and Goodman, C.S., Genetic analysis of Laminin A in Drosophila: extracellular matrix containing laminin A is required for ocellar axon pathfinding, *Development*, 122 (1996) 2611-21.

[54] Geuna, S., Tos, P., Battiston, B. and Giacobini-Robecchi, M.G., Bridging peripheral nerve defects with muscle-vein combined guides, *Neurol Res*, 26 (2004) 139-44.

[55] Glasby, M.A., Gschmeissner, S.G., Hitchcock, R.J., and Huang, C.L, The dependence of nerve regeneration through muscle grafts in the rat on the availability an dorientation of basement membrane, *J. Neurocytol.*, 15 (1986) 497-510.

[56] Glazner, G.W., Lupien, S., Miller, J.A. and Ishii, D.N., Insulin-like growth factor II increases the rate of sciatic nerve regeneration in rats, *Neuroscience*, 54 (1993) 791-7.

[57] Gluck, T., Ueber Neuroplastik auf dem Wege der Transplantation, *Arch Klin Chir*, 25 (1880) 606-616.

[58] Gospodarowicz, D., Ferrara, N., Schweigerer, L. and Neufeld, G., Structural characterization and biological functions of fibroblast growth factor, *Endocr Rev*, 8 (1987) 95-114.

[59] Griffin, C.G. and Letourneau, P.C., Rapid retraction of neurites by sensory neurons in response to increased concentrations of nerve growth factor, *J Cell Biol*, 86 (1980) 156-61.

[60] Groves, M.L., McKeon, R., Werner, E., Nagarsheth, M., Meador, W. and English, A.W., Axon regeneration in peripheral nerves is enhanced by proteoglycan degradation, *Exp Neurol*, 195 (2005) 278-292.

[61] Guenard, V., Kleitman, N., Morrissey, T.K., Bunge, R.P. and Aebischer, P., Syngeneic Schwann cells derived from adult nerves seeded in semipermeable

guidance channels enhance peripheral nerve regeneration, *J Neurosci*, 12 (1992) 3310-20.

[62] Gulati, A.K., Peripheral nerve regeneration through short- and long-term degenerated nerve transplants., *Brain Res*, 742 (1996) 265-270.

[63] Gundersen, R.W. and Barrett, J.N., Neuronal chemotaxis: chick dorsal-root axons turn toward high concentrations of nerve growth factor, *Science*, 206 (1979) 1079-80.

[64] Gundersen, R.W. and Barrett, J.N., Characterization of the turning response of dorsal root neurites toward nerve growth factor, *J Cell Biol*, 87 (1980) 546-54.

[65] Hadlock, T., Sundback, C., Hunter, D., Cheney, M. and Vacanti, J.P., A polymer foam conduit seeded with Schwann cells promotes guided peripheral nerve regeneration, *Tissue Eng*, 6 (2000) 119-27.

[66] Halfter, W., The behavior of optic axons on substrate gradients of retinal basal lamina proteins and merosin, *J Neurosci*, 16 (1996) 4389-401.

[67] Hall, S.M., The effect of inhibiting Schwann cell mitosis on the re-innervation of acellular autografts in the peripheral nervous system of the mouse, *Neuropathol Appl Neurobiol*, 12 (1986) 401-14.

[68] Heumann, R., Korsching, S., Bandtlow, C. and Thoenen, H., Changes of nerve growth factor synthesis in nonneuronal cells in response to sciatic nerve transection, *J Cell Biol*, 104 (1987) 1623-31.

[69] Hoffman, D., Wahlberg, L. and Aebischer, P., NGF released from a polymer matrix prevents loss of ChAT expression in basal forebrain neurons following a fimbria-fornix lesion, *Exp Neurol*, 110 (1990) 39-44.

[70] http://www.axogeninc.com, AxoGen Inc., *Vol. 2006*, AxoGen Inc, 2006.

[71] Hudson, T.W., Evans, G.R. and Schmidt, C.E., Engineering strategies for peripheral nerve repair, *Orthop Clin North Am*, 31 (2000) 485-98.

[72] Hynes, R.O., Integrins: a family of cell surface receptors, *Cell*, 48 (1987) 549-54.

[73] Ide, C., Peripheral nerve regeneration, *Neurosci Res*, 25 (1996) 101-21.

[74] Jansen, K., van der Werff, J.F., van Wachem, P.B., Nicolai, J.P., de Leij, L.F. and van Luyn, M.J., A hyaluronan-based nerve guide: in vitro cytotoxicity, subcutaneous tissue reactions, and degradation in the rat, *Biomaterials*, 25 (2004) 483-9.

[75] Jing, S., Tapley, P. and Barbacid, M., Nerve growth factor mediates signal transduction through trk homodimer receptors, *Neuron*, 9 (1992) 1067-79.

[76] Kam, L., Shain, W., Turner, J.N. and Bizios, R., Axonal outgrowth of hippocampal neurons on micro-scale networks of polylysine-conjugated laminin, *Biomaterials*, 22 (2001) 1049-54.

[77] Kapur, T.A. and Shoichet, M.S., Immobilized concentration gradients of nerve growth factor guide neurite outgrowth, *J Biomed Mater Res A*, 68 (2004) 235-43.

[78] Keilhoff, G., Pratsch, F., Wolf, G. and Fansa, H., Bridging extra large defects of peripheral nerves: possibilities and limitations of alternative biological grafts from acellular muscle and Schwann cells, *Tissue Eng*, 11 (2005) 1004-14.

[79] Kendal, E., Schwartz, J. and Jessal, T.M., The induction and patterning of nervous system., *Principles of Neuroscience*, McGraw-Hill, 2000, pp. 1019-1040.

[80] Kerkhoff, H. and Jennekens, F.G., Peripheral nerve lesions: the neuropharmacological outlook, *Clin Neurol Neurosurg*, 95 Suppl (1993) S103-8.

[81] Keynes, R. and Cook, G.M., Axon guidance molecules, *Cell*, 83 (1995) 161-9.

[82] Kirk, E.G. and D., L., Fascial tubulization in the repair of nerve defects, *JAMA*, 65 (1915) 486-492.

[83] Kiyotani, T., Nakamura, T., Shimuzu, Y., Endo, K., Experimental study of nerve regeneration in a biodegradable tube made from collagen and polyglycolic acid, *ASAIO J.*, 41 (1995) M657-661.

[84] Kleinman, H.K., Ogle, R.C., Cannon, F.B., Little, C.D., Sweeney, T.M. and Luckenbill-Edds, L., Laminin receptors for neurite formation, *Proc Natl Acad Sci U S A*, 85 (1988) 1282-6.

[85] Kline, D.G., Kim, D., Midha, R., Harsh, C. and Tiel, R., Management and results of sciatic nerve injuries: a 24-year experience, *J Neurosurgery*, 89 (1998) 13-23.

[86] Krause, T.L. and Bittner, G.D., Rapid morphological fusion of severed myelinated axons by polyethylene glycol, *Proc Natl Acad Sci U S A*, 87 (1990) 1471-5.

[87] Krewson, C.E., Klarman, M.L. and Saltzman, W.M., Distribution of nerve growth factor following direct delivery to brain interstitium, *Brain Res*, 680 (1995) 196-206.

[88] Kuffler, D.P., Isolated satellite cells of a peripheral nerve direct the growth of regenerating frog axons, *J Comp Neurol*, 249 (1986) 57-64.

[89] Labrador, R.O., Buti, M. and Navarro, X., Peripheral nerve repair: role of agarose matrix density on functional recovery, *Neuroreport*, 6 (1995) 2022-6.

[90] Labrador, R.O., Buti, M. and Navarro, X., Influence of collagen and laminin gels concentration on nerve regeneration after resection and tube repair, *Exp Neurol*, 149 (1998) 243-52.

[91] Lawson, G.M. and Glasby, M.A., Peripheral nerve reconstruction using freeze-thawed muscle grafts: a comparison with group fascicular nerve grafts in a large animal model, *J R Coll Surg Edinb.*, 43 (1998) 295-302.

[92] Lee, A.C., Yu, V.M., Lowe, J.B., 3rd, Brenner, M.J., Hunter, D.A., Mackinnon, S.E. and Sakiyama-Elbert, S.E., Controlled release of nerve growth factor enhances sciatic nerve regeneration, *Exp Neurol*, 184 (2003) 295-303.

[93] Letourneau, P.C., Chemotactic response of nerve fiber elongation to nerve growth factor, *Dev Biol*, 66 (1978) 183-96.

[94] Letourneau, P.C., Condic, M.L. and Snow, D.M., Interactions of developing neurons with the extracellular matrix, *J Neurosci*, 14 (1994) 915-28.

[95] Levi-Montalcini, R., The nerve growth factor 35 years later, *Science*, 237 (1987) 1154-62.

[96] Levi, A.D., Sonntag, V.K., Dickman, C., Mather, J., Li, R.H., Cordoba, S.C., Bichard, B. and Berens, M., The role of cultured Schwann cell grafts in the repair of gaps within the peripheral nervous system of primates, *Exp. Neurol.*, 143 (1997) 25-36.

[97] Li, Y., Decherchi, P. and Raisman, G., Transplantation of olfactory ensheathing cells into spinal cord lesions restores breathing and climbing, *J Neurosci*, 23 (2003) 727-31.

[98] Liesi, P., Do neurons in the vertebrate CNS migrate on laminin?, *Embo J*, 4 (1985) 1163-70.

[99] Liesi, P., Hager, G., Dodt, H.U., Seppala, I. and Zieglgansberger, W., Domain-specific antibodies against the B2 chain of laminin inhibit neuronal migration in the neonatal rat cerebellum, *J Neurosci Res*, 40 (1995) 199-206.

[100] Liuzzi, F.J. and Tedeschi, B., Peripheral nerve regeneration, *Neurosurg Clin N Am*, 2 (1991) 31-42.

[101] Luckenbill-Edds, L., Laminin and the mechanism of neuronal outgrowth, *Brain Research Reviews*, 23 (1997) 1-27.

[102] Lundborg, G., *Nerve Injury and Repair*, Longman Group UK, New York, 1988.

[103] Lundborg, G., Dahlin, L.B., Danielsen, N., Gelberman, R.H., Longo, F.M., Powell, H.C. and Varon, S., Nerve regeneration in silicone chambers: influence of gap length and of distal stump components, *Exp Neurol*, 76 (1982) 361-75.

[104] Mackinnon, S.E., Dellon, A.L., Clinical nerve reconstruction with a bioabsorbable polyglycolic acid tube, *Plast. Reconstr. Surg.*, 85 (1990) 419-424.

[105] Madison, R.D., Da Silva, C.F. and Dikkes, P., Entubulation repair with protein additives increases the maximum nerve gap distance successfully bridged with tubular prostheses, *Brain Res*, 447 (1988) 325-34.

[106] Martini, R., Expression and functional roles of neural cell surface molecules and extracellular matrix components during development and regeneration of peripheral nerves, *J Neurocytol*, 23 (1994) 1-28.

[107] Matsumoto, K., Ohnishi, K., Kiyotani, T., Sekine, T., Ueda, H., Nakamura, T., Endo, K. and Shimizu, Y., Peripheral nerve regeneration across an 80-mm gap bridged by a polyglycolic acid (PGA)-collagen tube filled with laminin-coated collagen fibers: a histological and electrophysiological evaluation of regenerated nerves, *Brain Res*, 868 (2000) 315-28.

[108] Matsuzawa, M., Tokumitsu, S., Knoll, W. and Leisi, P., Molecular gradeint along the axon pathway is not required for directional axon growth, *J Neurosci Res*, 53 (1998) 114-124.

[109] McCormack, M.L., Goddard, M., Guenard, V. and Aebischer, P., Comparison of dorsal and ventral spinal root regeneration through semipermeable guidance channels, *J Comp Neurol*, 313 (1991) 449-56.

[110] McKenna, M.P. and Raper, J.A., Growth cone behavior on gradients of substratum bound laminin, *Dev Biol*, 130 (1988) 232-236.

[111] McLoon, S.C., McLoon, L.K., Palm, S.L. and Furcht, L.T., Transient expression of laminin in the optic nerve of the developing rat, *J Neurosci*, 8 (1988) 1981-90.

[112] Meek, M.F., Varejo, A.S., Geuna, S., Muscle grafts and alternatives for nerve repair, *J. Oral Maxillofac. Surg.*, 60 (2002) 1095-96.

[113] Meilander, N.J., Yu, X., Ziats, N.P. and Bellamkonda, R.V., Lipid-based microtubular drug delivery vehicles, *J Control Release*, 71 (2001) 141-52.

[114] Meyer, R.S., Abrams, R.A., Botte, M.J., Davey, J.P. and Bodine-Fowler, S.C., Functional recovery following neurorrhaphy of the rat sciatic nerve by epineurial repair compared with tubulization, *J Orthop Res*, 15 (1997) 664-9.

[115] Midha, R., Nag, S., Munro, C.A. and Ang, L.C., Differential response of sensory and motor axons in nerve allografts after withdrawal of immunosuppressive therapy, *J Neurosurg*, 94 (2001) 102-10.

[116] Millesi, H., Meissl, G. and Berger, A., The interfascicular nerve-grafting of the median and ulnar nerves, *J Bone Joint Surg Am.*, 54 (1972) 7727-7750.

[117] Nakahara, Y., Gage, F.H. and Tuszynski, M.H., Grafts of fibroblasts genetically modified to secrete NGF, BDNF, NT-3, or basic FGF elicit differential responses in the adult spinal cord, *Cell Transplant*, 5 (1996) 191-204.

[118] Pagnotta, A., Tos, P., Fornaro, M., Gigante, A., Geuna, S., Neurtrophins and their receptors in early axonal regeneration along muscle-vein-combined grafts, *Microsurgery*, 22 (2002) 300-3.

[119] Payr, E., Beitrage zur Technik der Blutgefass und Nervennaht nebst Mittheilungen uber die Vervendung eines resorbibaren Metalles in der Chirurgie, *Arch Klin Chir*, 62 (1900) 67.

[120] Powell, S.K. and Kleinman, H.K., Neuronal laminins and their cellular receptors, *Int J Biochem Cell Biol*, 29 (1997) 401-14.

[121] Raivich, G. and Kreutzberg, G.W., Peripheral nerve regeneration: role of growth factors and their receptors, *Int J Dev Neurosci*, 11 (1993) 311-24.

[122] Reichardt, L.F. and Tomaselli, K.J., Extracellular matrix molecules and their receptors: functions in neural development, *Annu Rev Neurosci*, 14 (1991) 531-70.

[123] Rich, K.M., Luszczynski, J.R., Osborne, P.A. and Johnson, E.M., Jr., Nerve growth factor protects adult sensory neurons from cell death and atrophy caused by nerve injury, *J Neurocytol*, 16 (1987) 261-8.

[124] Rivas, R.J., Burmeister, D.W. and Goldberg, D.J., Rapid effects of laminin on the growth cone, *Neuron*, 8 (1992) 107-115.

[125] Rogers, S.L., Edson, K.J., Letourneau, P.C. and McLoon, S.C., Distribution of laminin in the developing peripheral nervous system of the chick, *Dev Biol*, 113 (1986) 429-35.

[126] Rogers, S.L., Letourneau, P.C., Palm, S.L., McCarthy, J. and Furcht, L.T., Neurite extension by central and peripheral nervous system neurons in response to substratun-bound fibronectin and laminin, *Dev Biol*, 98 (1983) 212-220.

[127] Rudolph, A.S., Stilwell, G., Cliff, R.O., Kahn, B., Spargo, B.J., Rollwagen, F. and Monroy, R.L., Biocompatibility of lipid microcylinders: effect on cell growth and antigen presentation in culture, *Biomaterials*, 13 (1992) 1085-92.

[128] Sakiyama-Elbert, S.E. and Hubbell, J.A., Development of fibrin derivatives for controlled release of heparin-binding growth factors, *J Control Release*, 65 (2000) 389-402.

[129] Saneinejad, S. and Shoichet, M.S., Patterned glass surfaces direct cell adhesion and process outgrowth of primary neurons of the central nervous system, *J Biomed Mater Res*, 42 (1998) 13-9.

[130] Satou, T., Nishida, S., Hiruma, S., Tanji, K., Takahashi, M., Fujita, S., Mizuhara, Y., Akai, F. and Hashimoto, S., A morphological study on the effects of collagen gel matrix on regeneration of severed rat sciatic nerve in silicone tubes, *Acta Pathol Jpn*, 36 (1986) 199-208.

[131] Schanen-King, C., Nel, A., Williams, L.K. and Landreth, G., Nerve growth factor stimulates the tyrosine phosphorylation of MAP2 kinase in PC12 cells, *Neuron*, 6 (1991) 915-22.

[132] Schmidt, C.E. and Leach, J.B., Neural Tissue Engineering: Strategies for repair and regeneration, *Annu. Rev. Biomed. Eng* (2003) 293-347.

[133] Seckel, B.R., Jones, D., Hekimian, K.J., Wang, K.K., Chakalis, D.P. and Costas, P.D., Hyaluronic acid through a new injectable nerve guide delivery system enhances peripheral nerve regeneration in the rat, *J Neurosci Res*, 40 (1995) 318-24.

[134] Sendtner, M., Holtmann, B., Kolbeck, R., Thoenen, H. and Barde, Y.A., Brain-derived neurotrophic factor prevents the death of motoneurons in newborn rats after nerve section, *Nature*, 360 (1992) 757-9.

[135] Song, H. and Poo, M., The cell biology of neuronal navigation, *Nat Cell Biol*, 3 (2001) E81-8.

[136] Spargo, B.J., Cliff, R.O., Rollwagen, F.M. and Rudolph, A.S., Controlled release of transforming growth factor-beta from lipid-based microcylinders, *J Microencapsul*, 12 (1995) 247-54.

[137] Strassman, R.J., Letourneau, P.C. and Wessells, N.K., Elongation of axons in an agar matrix that does not support cell locomotion, *Exp Cell Res*, 81 (1973) 482-7.

[138] Strauch, B., Rodriguez, D.M., Diaz, J., Yu, H.L., Kaplan, G. and Weinstein, D.E., Autologous Schwann cells drive regeneration through a 6-cm autogenous venous nerve conduit, *J Reconstr Microsurg*, 17 (2001) 589-595.

[139] Suematsu, N., Tubulation of the peripheral nerve gap: Its history and possibility, *Microsurgery*, 10 (1989) 1-74.

[140] Sunderland, S., *Nerve Injuries and their Repair: A Critical Appraisal.*, Churchill Livingstone, New York, 1991.

[141] Suzuki, Y., Tanihara, M., Ohnishi, K., Suzuki, K., Endo, K. and Nishimura, Y., Cat peripheral nerve regeneration across 50 mm gap repaired with a novel nerve guide composed of freeze-dried alginate gel, *Neurosci Lett*, 259 (1999) 75-8.

[142] Terzis, J., Faibisoff, B. and Williams, B., The nerve gap: suture under tension vs. graft, *Plast Reconstr Surg*, 56 (1975) 166-70.

[143] Tessier-Lavigne, M. and Goodman, C.S., The molecular biology of axon guidance, *Science*, 274 (1996) 1123-33.

[144] Thanos, P.K., Okajima, S. and Terzis, J.K., Utrastructure and cellualr biology of nerve regeneration, *J. Reconstr. Microsurg.*, 14 (1998) 423-36.

[145] Thoenen, H., Barde, Y.A., Davies, A.M. and Johnson, J.E., Neurotrophic factors and neuronal death, *Ciba Found Symp*, 126 (1987) 82-95.

[146] Timple, R., Immunological studies on collagen. In G.N. Ramachandran and A.H. Reddi (Eds.), *Biochemistry of collagen*, Plenum Press, New York, 1976, pp. 319-375.

[147] Tong, X.J., Hirai, K., Shimada, H., Mizutani, Y., Izumi, T., Toda, N. and Yu, P., Sciatic nerve regeneration navigated by laminin-fibronectin double coated biodegradable collagen grafts in rats, *Brain Res*, 663 (1994) 155-62.

[148] Tresco, P., Tissue engineering strategies for nervous system repair, *Progress in Brain Research*, 128 (2000) 349-363.

[149] Uckun, F.M., Evans, W.E., Forsyth, C.J., Waddick, K.G., Ahlgren, L.T., Chelstrom, L.M., Burkhardt, A., Bolen, J. and Myers, D.E., Biotherapy of B-cell precursor leukemia by targeting genistein to CD19-associated tyrosine kinases, *Science*, 267 (1995) 886-91.

[150] Uzman, B.G., Villegas, G.M., Mouse sciatic nerve regeneration through semi-permeable tubes: a quantitative model, *J Neurosci*, 9 (1983) 325-338.

[151] Valentini, R.F. and Aebischer, P., Strategies for the engineering of peripheral nervous tissue regeneration. In L.R. Lanza RP, Chick WL (Ed.), *Principles of Tissue Engineering*, R.G.Landes Company, Austin, 1997, pp. 671-684.

[152] Valentini, R.F., Aebischer, P., Winn, S.R. and Galletti, P.M., Collagen- and laminin-containing gels impede peripheral nerve regeneration through semipermeable nerve guidance channels, *Exp Neurol*, 98 (1987) 350-6.

[153] Valentini, R.F., Sabatini, A.M., Dario, P. and Aebischer, P., Polymer electret guidance channels enhance peripheral nerve regeneration in mice, *Brain Res*, 480 (1989) 300-4.

[154] Varejao, A.S., Melo-Pinto, P., Meek, M.F., Filipe, V.M. and Bulas-Cruz, J., Methods for the experimental functional assessment of rat sciatic nerve regeneration, *Neurol Res*, 26 (2004) 186-94.

[155] Walton, R.L., Brown, R.E., Matory, W.E. Jr, Borah, G.L., Dolph, J.L., Autogenous vein graft repair of digital nerve defects in the finger: a retrospective clinical study, *Plast. Reconstr. Surg.*, 84 (1989) 944-49.

[156] Wang, G.Y., Hirai, K. and Shimada, H., The role of laminin, a component of Schwann cell basal lamina, in rat sciatic nerve regeneration within antiserum-treated nerve grafts, *Brain Res*, 570 (1992a) 116-25.

[157] Wang, G.Y., Hirai, K., Shimada, H., Taji, S. and Zhong, S.Z., Behavior of axons, Schwann cells and perineurial cells in nerve regeneration within transplanted nerve grafts: effects of anti-laminin and anti-fibronectin antisera, *Brain Res*, 583 (1992b) 216-26.

[158] Wang, K.K., Costas, P.D., Bryan, D.J., Jones, D.S. and Seckel, B.R., Inside-out vein graft promotes improved nerve regeneration in rats, *J Reconstr Microsurg*, 14 (1993) 608-618.

[159] Wang, K.K., Nemeth, I.R., Seckel, B.R., Chakalis-Haley, D.P., Swann, D.A., Kuo, J.W., Bryan, D.J. and Cetrulo, C.L., Jr., Hyaluronic acid enhances peripheral nerve regeneration in vivo, *Microsurgery*, 18 (1998) 270-5.

[160] Weiss, P. and Taylor, A.C., Guides for nerve regeneration across nerve gaps, *J Neurosurg*, 3 (1946) 275-82.

[161] Williams, L.R., Longo, F.M., Powell, H.C., Lundborg, G. and Varon, S., Spatial-temporal progress of peripheral nerve regeneration within a silicone chamber: parameters for a bioassay, *J Comp Neurol*, 218 (1983) 460-70.

[162] Xu, X.M., Zhang, S.X., Li, H., Aebischer, P., and Bunge, M.B., Regrowth of axons into distal spinal cord through a Schwann cell seeded mini-channel implanted into hemisected adult rat spinal cord, *Eur. J. Neurosci.*, 11 (1999) 1723-1740.

[163] Yu, X. and Bellamkonda, R.V., Tissue-engineered scaffolds are effective alternatives to autografts for bridging peripheral nerve gaps, *Tissue Eng*, 9 (2003) 421-30.

[164] Yu, X., Dillon, G.P. and Bellamkonda, R.B., A laminin and nerve growth factor-laden three-dimensional scaffold for enhanced neurite extension, *Tissue Eng*, 5 (1999) 291-304.

[165] Zhang, F., Blain, B., Beck, J., Zhang, J., Chen, Z., Chen, Z.W. and Leineaweaver, W.C., Autogenous venous graft with one-stage prepared Schwann cells as a conduit for repair of long segmental nerve defects, *J Reconstr Microsurg*, 18 (2002) 295-300.

[166] Zhao, Q., Dahlin, L.B., Kanje, M. and Lundborg, G., Specificity of muscle reinnervation following repair of the transected sciatic nerve. A comparative study of different repair techniques in the rat, *J Hand Surg [Br]*, 17 (1992) 257-61.

[167] Zhou, F.C. and Azmitia, E.C., Laminin directs and facilitates migration and fiber growth of transplanted serotonin and norepinephrine neurons in adult brain, *Prog Brain Res*, 78 (1988) 413-26.

[168] Zuo, J., Neubauer, D., Graham, J., Krekoski, C.A., Ferguson, T.A. and Muir, D., Regeneration of axons after nerve transection repair is enhanced by degradation of chondroitin sulfate proteoglycan, *Exp Neurol*, 176 (2002) 221-8.

Chapter 4: **CONCLUSIONS AND PERSPECTIVES**

4.1. Conclusions

The goals of this thesis were: 1) to develop novel three-dimensional (3D) anisotropic scaffolds to promote enhanced neurite extension *in vitro*; and 2) to develop 3D anisotropic scaffolds with gradients of laminin-1 (LN-1) and nerve growth factor (NGF) to bridge long nerve gap in rats. There are 3 main achievements: 1) 3D agarose hydrogel scaffolds with immobilized gradients of ECM protein, LN-1, and diffuse gradient of neurotrophic factor, NGF, were synthesized using diffusion technique and photochemical immobilization. 2) We have demonstrated that anisotropic scaffolds with gradients of LN-1 promote enhanced neurite extension from chick dorsal root ganglia (DRG), as compared to isotropic scaffolds. 3) 3D tissue-engineered agarose scaffolds with gradients of both ECM protein (LN-1) and neurotrophic factor (NGF) successfully promote nerve regeneration to bridge a challenging 20-mm nerve gap in rats; where as isotropic scaffolds were unsuccessful in doing so. This model can be used to design 3D scaffolds with gradients for more than one molecule for *in vitro* and *in vivo* studies to bridge long nerve defects.

4.1.1. Anisotropic Scaffolds Promote Enhanced Neurite Extension *In Vitro*

3D Agarose hydrogel scaffolds with gradients of LN-1 were synthesized to promote enhanced neurite extension from chick DRG. A technique to generate diffuse gradients of NGF in 3D scaffolds has been previously described by Cao et al.[1]. However, to promote regeneration across long nerve gaps, immobilized gradients may be more sustainable than diffuse gradients over the period of several weeks during which peripheral nervous system (PNS) regeneration occurs. This study presents a technique to fabricate anisotropic 3D scaffolds with immobilized gradients of LN-1 and quantitatively evaluates the influence of these scaffolds on chick DRG neurite extension. Isotropic LN-1 scaffold and plain agarose scaffold (no LN-1) were used as controls. The results demonstrate that a particular concentration gradient (0.017 μg/ml/mm) of LN-1 promotes faster neurite extension than the highest neurite growth rate observed with isotropic LN-1 concentration (0.5 μg/ml). This suggests that gradients of LN-1 may be a more important parameter to optimize for maximal neurite extension rather than the absolute concentrations of isotropically distributed LN-1. When DRGs were cultured in anisotropic LN-1 scaffolds with a steep slope of LN-1 gradient (0.121 μg/ml/mm or 241% concentration difference across 2.5 mm scaffold), there was no difference in neurite growth rate "up the gradient", "down the gradient" and "perpendicular to gradient". When the steepness of the gradient was reduced to 0.051 μg/ml/mm (103.3% concentration difference across 2.5 mm scaffold), there was difference between neurite growth rate "up the gradient" and "down the gradient" (p-value < 0.05). When the steepness of the gradient was further reduced to 0.017 μg/ml/mm (or 35% concentration difference across 2.5 mm scaffold), there was even more difference between neurite growth "up the gradient" and "down the gradient" (p-value < 0.01). In the anisotropic scaffolds used in our studies, the extending neurites were exposed to either increasing or decreasing gradient of LN-1. The neurites that were exposed to decreasing gradient did not try to reorient themselves "up the gradient", but continued to grow "down the gradient". Similarly, neurites exposed to increasing gradient of LN-1 did not try to reorient themselves "down the gradient".

In summary, a novel technique to fabricate and characterize anisotropic LN-1 containing hydrogel scaffolds was established. The results clearly demonstrate that scaffolds with gradients of LN-1 promote faster neurite growth rate than maximal growth rate possible in isotropic LN-1 scaffolds. The results also demonstrate that there exists an optimal 'slope' for the gradient of LN-1, with the lowest slope in the range tested being the most optimal for the highest neurite growth rate. Although some studies have been done to understand how neural cells regulate their integrin expression in response to different isotropic concentrations of LN [2], in future studies it would be interesting to investigate how neurites adapt to changing concentrations of LN as in the case of gradients. Such anisotropic scaffolds, with directional cues, are likely to enhance peripheral nerve regeneration *in vivo* when presented across nerve gaps inside carrier polymer nerve guidance channels.

4.1.2. Bioengineered Scaffolds Promote Enhanced Nerve Regeneration in Rodents

To simulate *in vivo* conditions for axonal guidance and regeneration, a novel technique was designed to synthesize gradients of ECM proteins and neurotrophic factors. Here we show that anisotropic agarose gel scaffolds with gradients of LN-1 and NGF promote enhanced axonal regeneration, as compared to isotropic scaffolds with uniform concentrations of LN-1 and NGF, across a challenging 20-mm nerve gap in rats. In the nerve injury model studied here, regeneration is limited and does not occur unless the nerve guidance channels have gradients of both LN-1 and NGF (step- or continuous-).

Axonal regeneration was evaluated by histological analysis. Anisotropic scaffolds with a gradient of either LN-1 or NGF, maintaining a uniform concentration for the other component, did not promote axonal regeneration across the challenging 20-mm nerve gap. Although there was enhanced axonal regeneration in the anisotropic scaffolds, it did not result in enhanced motor function recovery as measured by walking-track analysis, relative gastrocnemius muscle weight and electrophysiology. Nerve grafts resulted in better functional outcome than anisotropic scaffolds. However, optimization of LN-1 and NGF concentration gradients; and longer duration of experimental study might lead to anisotropic scaffolds whose performance is comparable to nerve grafts.

Anisotropic scaffolds were synthesized by first making gradients of LN-1 in agarose gel by diffusion and then immobilizing the protein by photochemical cross-linking. Immobilization of LN-1 mimics *in vivo* conditions where LN-1 is tethered to ECM. Gradients of NGF were designed using slow-release LMTs, which mimics *in vivo* conditions where neurotrophins are released by target organs and taken up by nerve ends. The NGF gradient was always a continuous gradient due to the diffusion of NGF from LMTs. However, the LN-1 gradients were of two-kind, a step-gradient and a continuous-gradient. The step- and continuous-gradient scaffolds of LN-1 were designed such that the average LN-1 concentration in both conditions are similar. However, the actual concentration of LN-1 that the regenerating axons sense dynamically in each condition was different due to differing gradient profiles. The LN-1 concentration range selected for making LN-1 step-gradients was based on the maximum concentration of LN-1 that could be immobilized. However, this concentration range may not be the optimal range.

A more thorough study will be needed to determine the optimal LN-1 concentration range for nerve regeneration.

In summary, the *in vivo* work presents a novel technique to synthesize anisotropic scaffolds with gradients of ECM proteins and neurotrophic factors. These anisotropic scaffolds were successful in bridging a challenging 20-mm nerve gap while isotropic scaffolds were not successful. Although anisotropic scaffolds were comparable to nerve grafts in histological analysis, nerve grafts resulted in superior functional outcome. Optimization of LN-1 and NGF concentration in the anisotropic scaffold might improve its performance to match the nerve grafts.

4.2. Perspectives

4.2.1. Future studies related to development of *in vitro* model

In this study, LN-1 gradients with three different slopes were tested for promoting neurite extension from chick DRG. The anisotropic scaffold with LN-1 gradient of 0.017 μg/ml/mm showed the fastest neurite extension. However, a more thorough study with different slopes of LN-1, with different concentration ranges needs to be performed to determine the optimal concentration range and slope of LN-1 gradient. Anisotropic scaffolds with different slopes and concentration ranges of LN-1 gradients can be synthesized by changing the concentration of LN-1 solution used and the time-duration of diffusion process.

Although anisotropic scaffolds promote faster neurite extension than isotropic scaffolds, the exact mechanism is unknown. This could be due to regulation of receptors for LN-1 by chick DRG. The study conducted by Condic et al. demonstrated that neurite extension from chick DRG in LN-1 culture is mediated by mostly through α6β1 integrin receptors for LN-1. Chick DRG down-regulate receptors for LN-1 when cultured in high concentrations of LN-1 and up-regulate the receptors when cultured in low concentrations of LN-1 [2]. However, it is unknown if this mechanism is true for gradients as well. It is possible that in the 0.017 μg/ml/mm LN-1 gradient scaffolds, DRGs express more receptors for LN-1 (at the growth cones) than in the 0.051 μg/ml/mm and the 0.121 μg/ml/mm gradient scaffolds. Substrate-cytoskeletal coupling in the growth cone could transmit mechanical forces to pull the growth cone or its contents toward the anchored adhesive site. Therefore, in the anisotropic LN-1 scaffolds with lower slopes, higher density of receptors might lead to faster neurite growth rates by haptotactic mechanism. To establish the exact mechanism, expression of levels of receptors for LN-1 need to be determined.

4.2.2. Future studies related to development of *in vivo* model

In this study, gradients (step- or continuous-) of LN-1 and NGF were successful in promoting regeneration across a challenging 20-mm nerve gap in rats. However, the concentration of LN-1 and NGF used in our study has not been optimized. The highest possible concentration of LN-1 and NGF was used in this model, which may not be the optimal concentration. A more thorough study involving anisotropic scaffolds with different concentrations of LN-1 and NGF would be needed to determine the optimal concentration and the optimal slope of gradients of LN-1 and NGF.

In the anisotropic scaffolds, majority of the regenerating axons in the anisotropic scaffolds were small-diameter (0-2 μm) in nature. This could be due to the NGF in our scaffolds for nerve regeneration. NGF mainly promotes the survival and regeneration of sensory nerves, which are small-diameter in nature. BDNF on the other hand promotes survival and growth of both sensory and motor neurons. Using BDNF in our scaffolds might lead to regeneration of large-diameter motor neurons and also better motor functional outcome.

Our anisotropic scaffolds mimic *in vivo* conditions in presenting gradients haptotactic molecule, LN-1, and chemotactic molecule, NGF. However, studies have shown that presence of oriented substrates, such as nanofilaments, to physically guide the regenerating axons can further improve nerve regeneration [3]. In future studies, incorporation of physical substrates in these anisotropic scaffolds might further improve its performance.

In this study we have used non-degradable polysulfone tubes to present the anisotropic scaffolds. In clinical studies, after nerve regeneration, a second surgery would be needed to remove the non-degradable polysulfone tube, which is undesirable. In future studies, this extra procedure could be avoided by using bioresorbable material for making the nerve guidance channels, such as polyglycolide (PGA), polylactide (PLLA), PGA/PLLA blends and polycaprolactone [4].

4.3. References

[1] Cao, X. and Shoichet, M.S., Defining the concentration gradient of nerve growth factor for guided neurite outgrowth, *Neuroscience*, 103 (2001) 831-40.

[2] Condic, M.L. and Letourneau, P.C., Ligand-induced changes in integrin expression regulate neuronal adhesion and neurite outgrowth, *Nature*, 389 (1997) 852-856.

[3] Kim, Y., Haftel, V.K., Kumar, S. and Bellamkonda, R.V., Oriented nanoscaffolds match the performance of autografts in facilitatting regeneration across long peripheral nerve gaps, *Nat Biotechnol. (submitted).*

[4] Valentini, R.F. and Aebischer, P., Strategies for the engineering of peripheral nervous tissue regeneration. In L.R. Lanza RP, Chick WL (Ed.), *Principles of Tissue Engineering*, R.G.Landes Company, Austin, 1997, pp. 671-684.

References

[1] Adams, D.N., Kao, E.Y., Hypolite, C.L., Distefano, M.D., Hu, W.S. and Letourneau, P.C., Growth cones turn and migrate up an immobilized gradient of the laminin IKVAV peptide, *J Neurobiol*, 62 (2005) 134-47.

[2] Aebischer, P., Guenard, V. and Brace, S., Peripheral nerve regeneration through blind-ended semipermeable guidance channels:effect of molecular weight cutoff, *J Neurosci*, 9 (1989a) 3590-3595.

[3] Aebischer, P., Guenard, V. and Valentini, R.F., The morphology of regenerating peripheral nerves is modulated by the surface microgeometry of polymeric guidance channels, *Brain Res*, 531 (1990) 211-8.

[4] Aebischer, P., Salessiotis, A.N. and Winn, S.R., Basic fibroblast growth factor released from synthetic guidance channels facilitates peripheral nerve regeneration across long nerve gaps, *J Neurosci Res*, 23 (1989) 282-9.

[5] Aebischer, P., Valentini, R.F., Dario, P., Domenici, C. and Galletti, P.M., Piezoelectric guidance channels enhance regeneration in the mouse sciatic nerve after axotomy, *Brain Res*, 436 (1987) 165-8.

[6] Aebischer, P., Valentini, R.F., Winn, S.R. and Galletti, P.M., The use of a semi-permeable tube as a guidance channel for a transected rabbit optic nerve, *Prog Brain Res*, 78 (1988) 599-603.

[7] Ahmed, Z., Brown, R.A., Ljungberg, C., Wiberg, M. and Terenghi, G., Nerve growth factor enhances peripheral nerve regeneration in non-human primates, *Scand J Plast Reconstr Surg Hand Surg*, 33 (1999) 393-401.

[8] Albert, E., Einige Operationen an Nerven, *Wien Med*, 26 (1885) 1285.

[9] Amoh, Y., Li, L., Campillo, R., Kawahara, K., Katsuoka, K., Penman, S. and Hoffman, R.M., Implanted hair follicle stem cells form Schwann cells that support repair of severed peripheral nerves, *Proc Natl Acad Sci U S A*, 102 (2005) 17734-8.

[10] Ansselin, A.D., Fink, T. and Davey, D.F., Peripheral nerve regeneration through nerve guides seeded with adult Schwann cells, *Neuropathol Appl Neurobiol*, 23 (1997) 387-98.

[11] Ansselin, A.D. and Hudson, A.R., Axonal nerve regeneration through peripheral nerve grafts: the effect of proxio-distal orientation, *Microsurgery*, 9 (1988) 103-110.

[12] Baier, H. and Bonhoeffer, F., Axon guidance by gradients of a target-derived component, *Science*, 255 (1992) 472-5.

[13] Barde, Y.A., Trophic factors and neuronal survival, *Neuron*, 2 (1989) 1525-34.

[14] Baron-van Evercooren, A., Kleinman, H.D., Ohno, S., Marangos, P., Schwartz, J.P. and Dubois-Dalcq, M.E., Nerv growth factor, laminin and fibronectin promote nerve growth in human fetal sensory ganglia cultures, *J Neurosci. Res.* (1982) 179-183.

[15] Battiston, B., Tos, P., Cushway, T.R. and Geuna, S., Nerve repair by means of vein filled with muscle grafts I. Clinical results, *Microsurgery*, 20 (2000) 32-36.

[16] Bellamkonda, R.V., Ranieri, J.P., Bouche, N. and Aebischer, P., Hydrogel-based three-dimensional matrix for neural cells, *J Biomed Mater Res*, 29 (1995) 663-71.

[17] Bonhoeffer, F. and Huf, J., In vitro experiments on axon guidance demonstrating an anterior-posterior gradient on the tectum, *Embo J*, 1 (1982) 427-31.

[18] Bonner, J. and O'Connor, T.P., The permissive cue laminin is essential for growth cone turning in vivo, *J Neurosci*, 21 (2001) 9782-91.

[19] Brandt, J., Dahlin, L.B. and Lundborg, G., Autologous tendons used as grafts for bridging peripheral nerve defects, *J Hand Surg [Br]*, 24 (1999) 284-90.

[20] Bronner-Fraser, M., Stern, C.D. and Fraser, S., Analysis of neural crest cell lineage and migration, *J Craniofac Genet Dev Biol*, 11 (1991) 214-22.

[21] Brunelli, G., Fontana, G., Jager, C., Bartolaminelli, P. and Franchini, A., Chemotactic arrangement of axons inside and distal to a venous graft, *J Reconstr Microsurg*, 3 (1987) 87-93.

[22] Bryan, D.J., Wang, K.-K. and Chakalis-Haley, D.P., Effect of Schwann cells in the enhancement of peripheral nerve regeneration, *J. Reconstr. Microsurg.*, 12 (1996) 439-446.

[23] Bunge, R.P., Expanding roles for the Schwann cell: ensheathment, myelination, trophism and regeneration, *Curr Opin Neurobiol*, 3 (1993) 805-9.

[24] Cai, J., Peng, X., Nelson, K.D., Eberhart, R. and Smith, G.M., Permeable guidance channels containing microfilament scaffolds enhance axon growth and maturation, *J Biomed Mater Res A*, 75 (2005) 374-86.

[25] Cajal, S.R.y., Degeneration and regeneration of the nervous system, *Hafner* (1928).

[26] Cao, X. and Shoichet, M.S., Defining the concentration gradient of nerve growth factor for guided neurite outgrowth, *Neuroscience*, 103 (2001) 831-40.

[27] Cao, X. and Shoichet, M.S., Investigating the synergistic effect of combined neurotrophic factor concentration gradients to guide axonal growth, *Neuroscience*, 122 (2003) 381-9.

[28] Ceballos, D., Navarro, X., Dubey, N., Wendelschafer-Crabb, G., Kennedy, W.R. and Tranquillo, R.T., Magnetically aligned collagen gel filling a collagen nerve guide improves peripheral nerve regeneration, *Exp Neurol*, 158 (1999) 290-300.

[29] Champion, S., Imhof, B.A., Savagner, P. and Thiery, J.P., The embryonic thymus produces chemotactic peptides involved in the homing of hemopoietic precursors, *Cell*, 44 (1986) 781-90.

[30] Chen, Y.S., Hsieh, C.L., Tsai, C.C., Chen, T.H., Cheng, W.C., Hu, C.L. and Yao, C.H., Peripheral nerve regeneration using silicone rubber chambers filled with collagen, laminin and fibronectin, *Biomaterials*, 21 (2000) 1541-7.

[31] Chiu, D.T., Janecka, I., Krizek, T.J., Wolff, M. and Lovelace, R.E., Autogenous vein graft as a conduit for nerve regeneration, *Surgery*, 91 (1982) 226-33.

[32] Chiu, D.T., Lovelace, R.E., Yu, L.T., Wolff, M., Stengel, S., Middleton, L., Janecka, I.P., and Krizek, T.J., Comparative electrophysiological evaluation of nerve grafts and autogenous vein grafts as nerve conduits: an experimental study, *J. Reconstr. Microsurg.*, 4 (1988) 303-309.

[33] Chiu, D.T. and Strauch, B., A prospective clinical evaluation of autogenous vein grafts used as a nerve conduit for distal sensory nerve defects of 3 cm or less, *Plast Reconstr Surg*, 86 (1990) 928-34.

[34] Choi, B.H., Zhu, S.J., Kim, B.Y., Huh, J.Y., Lee, S.H. and Jung, J.H., Transplantation of cultured bone marrow stromal cells to improve peripheral nerve regeneration, *Int J Oral Maxillofac Surg*, 34 (2005) 537-42.

[35] Choi, B.H., Zhu, S.J., Kim, S.H., Kim, B.Y., Huh, J.H., Lee, S.H. and Jung, J.H., Nerve repair using a vein graft filled with collagen gel, *J Reconstr Microsurg*, 21 (2005) 267-72.

[36] Clark, P., Britland, S. and Connolly, P., Growth cone guidance and neuron morphology on micropatterned laminin surfaces, *J Cell Sci*, 105 (Pt 1) (1993) 203-12.

[37] Condic, M.L. and Letourneau, P.C., Ligand-induced changes in integrin expression regulate neuronal adhesion and neurite outgrowth, *Nature*, 389 (1997) 852-856.

[38] Cornbrooks, C.J., Carey, D.J., McDonald, J.A., Timpl, R. and Bunge, R.P., In vivo and in vitro observations on laminin production by Schwann cells, *Proc Natl Acad Sci U S A*, 80 (1983) 3850-4.

[39] Crank, J., Diffusion in a plane sheet. *The mathematics of diffusion*, Oxford University Press, London, 1970, pp. 44-68.

[40] Cussler, E.L., Values of diffusion coefficients. *Diffusion, Mass transfer in fluid systems*, Cambridge University Press, New York, 1984, pp. 101-141.

[41] Dahlin, L.B. and Lundborg, G., Use of tubes in peripheral nerve repair, *Neurosurg Clin N Am*, 12 (2001) 341-52.

[42] Danielsen, N., Dahlin, L.B., Lee, Y.F. and Lundborg, G., Axonal growth in mesothelial chambers. The role of the distal nerve segment, *Scand J Plast Reconstr Surg*, 17 (1983) 119-25.

[43] de Medinaceli, L., Freed, W.J. and Wyatt, R.J., An index of the functional condition of rat sciatic nerve based on measurements made from walking tracks, *Exp Neurol*, 77 (1982) 634-43.

[44] Derby, A., Engleman, V.W., Frierdich, G.E., Neises, G., Rapp, S.R. and Roufa, D.G., Nerve growth factor facilitates regeneration across nerve gaps: morphological and behavioral studies in rat sciatic nerve, *Exp Neurol*, 119 (1993) 176-91.

[45] Dertinger, S.K., Jiang, X., Li, Z., Murthy, V.N. and Whitesides, G.M., Gradients of substrate-bound laminin orient axonal specification of neurons, *Proc Natl Acad Sci U S A*, 99 (2002) 12542-7.

[46] Dillon, G.P., Yu, X. and Bellamkonda, R.V., The polarity and magnitude of ambient charge influences three-dimensional neurite extension from DRGs, *J Biomed Mater Res*, 51 (2000) 510-9.

[47] Dillon, G.P., Yu, X., Sridharan, A., Ranieri, J.P. and Bellamkonda, R.V., The influence of physical structure and charge on neurite extension in a 3D hydrogel scaffold, *J Biomater Sci Polym Ed*, 9 (1998) 1049-69.

[48] Dodla, M.C. and Bellamkonda, R.V., Anisotropic scaffolds facilitate enhanced neurite extension in vitro, *J Biomed Mater Res A*, 78 (2006) 213-21.

[49] Dubey, N., Letourneau, P.C. and Tranquillo, R.T., Guided neurite elongation and schwann cell invasion into magnetically aligned collagen in simulated peripheral nerve regeneration, *Exp Neurol*, 158 (1999) 338-50.

[50] Dubey, N., Letourneau, P.C. and Tranquillo, R.T., Neuronal contact guidance in magnetically aligned fibrin gels: effect of variation in gel mechano-structural properties, *Biomaterials*, 22 (2001) 1065-75.

[51] Edelman, E.R., Mathiowitz, E., Langer, R. and Klagsbrun, M., Controlled and modulated release of basic fibroblast growth factor, *Biomaterials*, 12 (1991) 619-26.

[52] Edgar, D., Nerve growth factors and molecules of the extracellular matrix in neuronal development, *J Cell Sci Suppl*, 3 (1985) 107-13.

[53] English, A.W., Meador, W. and Carrasco, D.I., Neurotrophin-4/5 is required for the early growth of regenerating axons in peripheral nerves., *Eur. J. Neurosci.*, 21 (2005) 2624-34.

[54] Esch, T., Lemmon, V. and Banker, G., Local presentation of substrate molecules directs axon specification by cultured hippocampal neurons, *J Neurosci*, 19 (1999) 6417-6426.

[55] Evans, G.R., Peripheral nerve injury: a review and approach to tissue engineered constructs, *Anat Rec*, 263 (2001) 396-404.

[56] Evans, G.R.D., Challenges to nerve regeneration, *Seminars in Surgical Oncology*, 19 (2000) 312-318.

[57] Evans, P.J., MacKinnon, S.E., Midha, R., Wade, J.A., Hunter, D.A., Nakao, Y. and Hare, G.M., Regeneration across cold preserved peripheral nerve allografts, *Microsurgery*, 19 (1999) 115-27.

[58] Favaro, G., Bortolami, M.C., Cereser, S., Dona, M., Pastorello, A., Callegaro, L. and Fiori, M.G., Peripheral nerve regeneration through a novel bioresorbable nerve guide, *ASAIO Trans*, 36 (1990) M291-4.

[59] Fawcett, J.W.a.K., R.J., Peripheral nerve regeneration, *Ann. Rev. Neurosci.*, 13 (1990) 43-60.

[60] Ferrari, F., De Castro Rodrigues, A., Malvezzi, C.K., Dal Pai Silava, M. and Padvoni, C.R., Inside-out vs. standard vein graft to repair a sensory nerve in rats, *Anat Rec*, 256 (1999) 227-232.

[61] Forrester, W.C. and Garriga, G., Genes necessary for C. elegans cell and growth cone migrations, *Development*, 124 (1997) 1831-43.

[62] Fu, S.Y. and Gordon, T., The cellular and molecular basis of peripheral nerve regeneration, *Mol Neurobiol*, 14 (1997) 67-116.

[63] Fujimoto, E., Mizoguchi, A., Hanada, K., Yajima, M. and Ide, C., Basic fibroblast growth factor promotes extension of regenerating axons of peripheral nerve. In vivo experiments using a Schwann cell basal lamina tube model, *J Neurocytol*, 26 (1997) 511-28.

[64] Gallo, G., Lefcort, F.B. and Letourneau, P.C., The trkA receptor mediates growth cone turning toward a localized source of nerve growth factor, *J Neurosci*, 17 (1997) 5445-54.

[65] Garcia-Alonso, L., Fetter, R.D. and Goodman, C.S., Genetic analysis of Laminin A in Drosophila: extracellular matrix containing laminin A is required for ocellar axon pathfinding, *Development*, 122 (1996) 2611-21.

[66] Geuna, S., Tos, P., Battiston, B. and Giacobini-Robecchi, M.G., Bridging peripheral nerve defects with muscle-vein combined guides, *Neurol Res*, 26 (2004) 139-44.

[67] Glasby, M.A., Gschmeissner, S.G., Hitchcock, R.J., and Huang, C.L, The dependence of nerve regeneration through muscle grafts in the rat on the availability an dorientation of basement membrane, *J. Neurocytol.*, 15 (1986) 497-510.

[68] Glazner, G.W., Lupien, S., Miller, J.A. and Ishii, D.N., Insulin-like growth factor II increases the rate of sciatic nerve regeneration in rats, *Neuroscience*, 54 (1993) 791-7.

[69] Gluck, T., Ueber Neuroplastik auf dem Wege der Transplantation, *Arch Klin Chir*, 25 (1880) 606-616.

[70] Goodman, C.S., Mechanisms and molecules that control growth cone guidance, *Annu Rev Neurosci*, 19 (1996) 341-77.

[71] Gospodarowicz, D., Ferrara, N., Schweigerer, L. and Neufeld, G., Structural characterization and biological functions of fibroblast growth factor, *Endocr Rev*, 8 (1987) 95-114.

[72] Griffin, C.G. and Letourneau, P.C., Rapid retraction of neurites by sensory neurons in response to increased concentrations of nerve growth factor, *J Cell Biol*, 86 (1980) 156-61.

[73] Groves, M.L., McKeon, R., Werner, E., Nagarsheth, M., Meador, W. and English, A.W., Axon regeneration in peripheral nerves is enhanced by proteoglycan degradation, *Exp Neurol*, 195 (2005) 278-292.

[74] Guenard, V., Kleitman, N., Morrissey, T.K., Bunge, R.P. and Aebischer, P., Syngeneic Schwann cells derived from adult nerves seeded in semipermeable guidance channels enhance peripheral nerve regeneration, *J Neurosci*, 12 (1992) 3310-20.

[75] Gulati, A.K., Peripheral nerve regeneration through short- and long-term degenerated nerve transplants., *Brain Res*, 742 (1996) 265-270.

[76] Gundersen, R.W. and Barrett, J.N., Neuronal chemotaxis: chick dorsal-root axons turn toward high concentrations of nerve growth factor, *Science*, 206 (1979) 1079-80.

[77] Gundersen, R.W. and Barrett, J.N., Characterization of the turning response of dorsal root neurites toward nerve growth factor, *J Cell Biol*, 87 (1980) 546-54.

[78] Hadlock, T., Sundback, C., Hunter, D., Cheney, M. and Vacanti, J.P., A polymer foam conduit seeded with Schwann cells promotes guided peripheral nerve regeneration, *Tissue Eng*, 6 (2000) 119-27.

[79] Halfter, W., The behavior of optic axons on substrate gradients of retinal basal lamina proteins and merosin, *J Neurosci*, 16 (1996) 4389-401.

[80] Hall, S.M., The effect of inhibiting Schwann cell mitosis on the re-innervation of acellular autografts in the peripheral nervous system of the mouse, *Neuropathol Appl Neurobiol*, 12 (1986) 401-14.

[81] Heumann, R., Korsching, S., Bandtlow, C. and Thoenen, H., Changes of nerve growth factor synthesis in nonneuronal cells in response to sciatic nerve transection, *J Cell Biol*, 104 (1987) 1623-31.

[82] Hoffman, D., Wahlberg, L. and Aebischer, P., NGF released from a polymer matrix prevents loss of ChAT expression in basal forebrain neurons following a fimbria-fornix lesion, *Exp Neurol*, 110 (1990) 39-44.

[83] http://www.axogeninc.com, AxoGen Inc., October 10, 2006.

[84] Hudson, T.W., Evans, G.R. and Schmidt, C.E., Engineering strategies for peripheral nerve repair, *Orthop Clin North Am*, 31 (2000) 485-98.

[85] Hynes, R.O., Integrins: a family of cell surface receptors, *Cell*, 48 (1987) 549-54.

[86] Ide, C., Peripheral nerve regeneration, *Neurosci Res*, 25 (1996) 101-21.

[87] Jansen, K., van der Werff, J.F., van Wachem, P.B., Nicolai, J.P., de Leij, L.F. and van Luyn, M.J., A hyaluronan-based nerve guide: in vitro cytotoxicity,

subcutaneous tissue reactions, and degradation in the rat, *Biomaterials*, 25 (2004) 483-9.

[88] Jing, S., Tapley, P. and Barbacid, M., Nerve growth factor mediates signal transduction through trk homodimer receptors, *Neuron*, 9 (1992) 1067-79.

[89] Kam, L., Shain, W., Turner, J.N. and Bizios, R., Axonal outgrowth of hippocampal neurons on micro-scale networks of polylysine-conjugated laminin, *Biomaterials*, 22 (2001) 1049-54.

[90] Kapur, T.A. and Shoichet, M.S., Immobilized concentration gradients of nerve growth factor guide neurite outgrowth, *J Biomed Mater Res A*, 68 (2004) 235-43.

[91] Keilhoff, G., Pratsch, F., Wolf, G. and Fansa, H., Bridging extra large defects of peripheral nerves: possibilities and limitations of alternative biological grafts from acellular muscle and Schwann cells, *Tissue Eng*, 11 (2005) 1004-14.

[92] Kendal, E., Schwartz, J. and Jessal, T.M., The induction and patterning of nervous system., *Principles of Neuroscience*, McGraw-Hill, 2000, pp. 1019-1040.

[93] Kerkhoff, H. and Jennekens, F.G., Peripheral nerve lesions: the neuropharmacological outlook, *Clin Neurol Neurosurg*, 95 Suppl (1993) S103-8.

[94] Keynes, R. and Cook, G.M., Axon guidance molecules, *Cell*, 83 (1995) 161-9.

[95] Kim, Y., Haftel, V.K., Kumar, S. and Bellamkonda, R.V., Oriented nanoscaffolds match the performance of autografts in facilitatting regeneration across long peripheral nerve gaps, *Nat Biotechnol. (submitted)*.

[96] Kirk, E.G. and D., L., Fascial tubulization in the repair of nerve defects, *JAMA*, 65 (1915) 486-492.

[97] Kiyotani, T., Nakamura, T., Shimuzu, Y., Endo, K., Experimental study of nerve regeneration in a biodegradable tube made from collagen and polyglycolic acid, *ASAIO J.*, 41 (1995) M657-661.

[98] Kleinman, H.K., Ogle, R.C., Cannon, F.B., Little, C.D., Sweeney, T.M. and Luckenbill-Edds, L., Laminin receptors for neurite formation, *Proc Natl Acad Sci U S A*, 85 (1988) 1282-6.

[99] Kline, D.G., Kim, D., Midha, R., Harsh, C. and Tiel, R., Management and results of sciatic nerve injuries: a 24-year experience, *J Neurosurgery*, 89 (1998) 13-23.

[100] Krause, T.L. and Bittner, G.D., Rapid morphological fusion of severed myelinated axons by polyethylene glycol, *Proc Natl Acad Sci U S A*, 87 (1990) 1471-5.

[101] Krewson, C.E., Klarman, M.L. and Saltzman, W.M., Distribution of nerve growth factor following direct delivery to brain interstitium, *Brain Res*, 680 (1995) 196-206.

[102] Kuffler, D.P., Isolated satellite cells of a peripheral nerve direct the growth of regenerating frog axons, *J Comp Neurol*, 249 (1986) 57-64.

[103] Kuhn, T.B., Williams, C.V., Dou, P. and Kater, S.B., Laminin directs growth cone navigation via two temporally and functionally distinct calcium signals, *J Neurosci*, 18 (1998) 184-94.

[104] Labrador, R.O., Buti, M. and Navarro, X., Peripheral nerve repair: role of agarose matrix density on functional recovery, *Neuroreport*, 6 (1995) 2022-6.

[105] Labrador, R.O., Buti, M. and Navarro, X., Influence of collagen and laminin gels concentration on nerve regeneration after resection and tube repair, *Exp Neurol*, 149 (1998) 243-52.

[106] Lawson, G.M. and Glasby, M.A., Peripheral nerve reconstruction using freeze-thawed muscle grafts: a comparison with group fascicular nerve grafts in a large animal model, *J R Coll Surg Edinb.*, 43 (1998) 295-302.

[107] Lee, A.C., Yu, V.M., Lowe, J.B., 3rd, Brenner, M.J., Hunter, D.A., Mackinnon, S.E. and Sakiyama-Elbert, S.E., Controlled release of nerve growth factor enhances sciatic nerve regeneration, *Exp Neurol*, 184 (2003) 295-303.

[108] Letourneau, P.C., Chemotactic response of nerve fiber elongation to nerve growth factor, *Dev Biol*, 66 (1978) 183-96.

[109] Letourneau, P.C., Condic, M.L. and Snow, D.M., Interactions of developing neurons with the extracellular matrix, *J Neurosci*, 14 (1994) 915-28.

[110] Levi-Montalcini, R., The nerve growth factor 35 years later, *Science*, 237 (1987) 1154-62.

[111] Levi, A.D., Sonntag, V.K., Dickman, C., Mather, J., Li, R.H., Cordoba, S.C., Bichard, B. and Berens, M., The role of cultured Schwann cell grafts in the repair of gaps within the peripheral nervous system of primates, *Exp. Neurol.*, 143 (1997) 25-36.

[112] Li, Y., Decherchi, P. and Raisman, G., Transplantation of olfactory ensheathing cells into spinal cord lesions restores breathing and climbing, *J Neurosci*, 23 (2003) 727-31.

[113] Liesi, P., Do neurons in the vertebrate CNS migrate on laminin?, *Embo J*, 4 (1985) 1163-70.

[114] Liesi, P., Hager, G., Dodt, H.U., Seppala, I. and Zieglgansberger, W., Domain-specific antibodies against the B2 chain of laminin inhibit neuronal migration in the neonatal rat cerebellum, *J Neurosci Res*, 40 (1995) 199-206.

[115] Liuzzi, F.J. and Tedeschi, B., Peripheral nerve regeneration, *Neurosurg Clin N Am*, 2 (1991) 31-42.

[116] Luckenbill-Edds, L., Laminin and the mechanism of neuronal outgrowth, *Brain Research Reviews*, 23 (1997) 1-27.

[117] Lundborg, G., *Nerve Injury and Repair*, Longman Group UK, New York, 1988.

[118] Lundborg, G., Dahlin, L.B., Danielsen, N., Gelberman, R.H., Longo, F.M., Powell, H.C. and Varon, S., Nerve regeneration in silicone chambers: influence of gap length and of distal stump components, *Exp Neurol*, 76 (1982) 361-75.

[119] Mackinnon, S.E., New directions in peripheral nerve surgery, *Ann Plastic Surg*, 22 (1989) 257-273.

[120] Mackinnon, S.E. and Dellon, A.L., Surgery of the Peripheral Nerve. Thieme Med. Publ., New York, 1988.

[121] Mackinnon, S.E., Dellon, A.L., Clinical nerve reconstruction with a bioabsorbable polyglycolic acid tube, *Plast. Reconstr. Surg.*, 85 (1990) 419-424.

[122] Madison, R.D., Da Silva, C.F. and Dikkes, P., Entubulation repair with protein additives increases the maximum nerve gap distance successfully bridged with tubular prostheses, *Brain Res*, 447 (1988) 325-34.

[123] Martini, R., Expression and functional roles of neural cell surface molecules and extracellular matrix components during development and regeneration of peripheral nerves, *J Neurocytol*, 23 (1994) 1-28.

[124] Matsumoto, K., Ohnishi, K., Kiyotani, T., Sekine, T., Ueda, H., Nakamura, T., Endo, K. and Shimizu, Y., Peripheral nerve regeneration across an 80-mm gap bridged by a polyglycolic acid (PGA)-collagen tube filled with laminin-coated collagen fibers: a histological and electrophysiological evaluation of regenerated nerves, *Brain Res*, 868 (2000) 315-28.

[125] Matsuzawa, M., Tokumitsu, S., Knoll, W. and Leisi, P., Molecular gradeint along the axon pathway is not required for directional axon growth, *J Neurosci Res*, 53 (1998) 114-124.

[126] McCormack, M.L., Goddard, M., Guenard, V. and Aebischer, P., Comparison of dorsal and ventral spinal root regeneration through semipermeable guidance channels, *J Comp Neurol*, 313 (1991) 449-56.

[127] McKenna, M.P. and Raper, J.A., Growth cone behavior on gradients of substratum bound laminin, *Dev Biol*, 130 (1988) 232-236.

[128] McLoon, S.C., McLoon, L.K., Palm, S.L. and Furcht, L.T., Transient expression of laminin in the optic nerve of the developing rat, *J Neurosci*, 8 (1988) 1981-90.

[129] Meek, M.F., Varejo, A.S., Geuna, S., Muscle grafts and alternatives for nerve repair, *J. Oral Maxillofac. Surg.*, 60 (2002) 1095-96.

[130] Meilander, N.J., Yu, X., Ziats, N.P. and Bellamkonda, R.V., Lipid-based microtubular drug delivery vehicles, *J Control Release*, 71 (2001) 141-52.

[131] Meyer, R.S., Abrams, R.A., Botte, M.J., Davey, J.P. and Bodine-Fowler, S.C., Functional recovery following neurorrhaphy of the rat sciatic nerve by epineurial repair compared with tubulization, *J Orthop Res*, 15 (1997) 664-9.

[132] Midha, R., Nag, S., Munro, C.A. and Ang, L.C., Differential response of sensory and motor axons in nerve allografts after withdrawal of immunosuppressive therapy, *J Neurosurg*, 94 (2001) 102-10.

[133] Millesi, H., Meissl, G. and Berger, A., The interfascicular nerve-grafting of the median and ulnar nerves, *J Bone Joint Surg Am.*, 54 (1972) 7727-7750.

[134] Millesi, H., Meissl, G. and Berger, A., Further experience with interfascicular grafting of the median, ulnar, and radial nerves, *J Bone Joint Surg Am*, 58 (1976) 209-18.

[135] Nakahara, Y., Gage, F.H. and Tuszynski, M.H., Grafts of fibroblasts genetically modified to secrete NGF, BDNF, NT-3, or basic FGF elicit differential responses in the adult spinal cord, *Cell Transplant*, 5 (1996) 191-204.

[136] Ngo, T.T., Waggoner, P.J., Romero, A.A., Nelson, K.D., Eberhart, R.C. and Smith, G.M., Poly(L-Lactide) microfilaments enhance peripheral nerve regeneration across extended nerve lesions, *J Neurosci Res*, 72 (2003) 227-38.

[137] Nichols, C.M., Brenner, M.J., Fox, I.K., Tung, T.H., Hunter, D.A., Rickman, S.R. and Mackinnon, S.E., Effects of motor versus sensory nerve grafts on peripheral nerve regeneration, *Exp Neurol*, 190 (2004) 347-55.

[138] Nobel J., M.C., Prasad VS and Midha R, Analysis of upper and lower extremity peripheral nerve injuries in a population of patients with multiple injuries, *J Neurotrauma*, 45 (1998) 116-122.

[139] Nunley JA, S.A., Sandow MJ and Urbaniak JR, Results of interfascicular nerve grafting for radial nerve lesions, *Microsurgery*, 17 (1996) 431-437.

[140] Pagnotta, A., Tos, P., Fornaro, M., Gigante, A., Geuna, S., Neurtrophins and their receptors in early axonal regeneration along muscle-vein-combined grafts, *Microsurgery*, 22 (2002) 300-3.

[141] Pandurangi, R.S., Karra, S.R., Katti, K.V., Kuntz, R.R. and Volkert, W.A., Chemistry of Bifunctional Photoprobes. 1. Perfluoroaryl Azido Functionalized Phosphorus Hydrazides as Novel Photoreactive Heterobifunctional Chelating Agents: High Efficiency Nitrene Insertion on Model Solvents and Proteins, *J Org Chem*, 62 (1997) 2798-2807.

[142] Pandurangi, R.S., Lusiak, P., Desai, S. and Kuntz, R.R., Chemistry of bifunctional photoprobes, *Bioorganic Chemistry*, 26 (1998) 201-212.

[143] Payr, E., Beitrage zur Technik der Blutgefass und Nervennaht nebst Mittheilungen uber die Vervendung eines resorbibaren Metalles in der Chirurgie, *Arch Klin Chir*, 62 (1900) 67.

[144] Powell, S.K. and Kleinman, H.K., Neuronal laminins and their cellular receptors, *Int J Biochem Cell Biol*, 29 (1997) 401-14.

[145] Properzi, F., Asher, R.A. and Fawcett, J.W., Chondroitin sulphate proteoglycans in the central nervous system: changes and synthesis after injury, *Biochem Soc Trans*, 31 (2003) 335-6.

[146] Raivich, G. and Kreutzberg, G.W., Peripheral nerve regeneration: role of growth factors and their receptors, *Int J Dev Neurosci*, 11 (1993) 311-24.

[147] Reichardt, L.F. and Tomaselli, K.J., Extracellular matrix molecules and their receptors: functions in neural development, *Annu Rev Neurosci*, 14 (1991) 531-70.

[148] Rich, K.M., Luszczynski, J.R., Osborne, P.A. and Johnson, E.M., Jr., Nerve growth factor protects adult sensory neurons from cell death and atrophy caused by nerve injury, *J Neurocytol*, 16 (1987) 261-8.

[149] Rivas, R.J., Burmeister, D.W. and Goldberg, D.J., Rapid effects of laminin on the growth cone, *Neuron*, 8 (1992) 107-115.

[150] Rogers, S.L., Edson, K.J., Letourneau, P.C. and McLoon, S.C., Distribution of laminin in the developing peripheral nervous system of the chick, *Dev Biol*, 113 (1986) 429-35.

[151] Rogers, S.L., Letourneau, P.C., Palm, S.L., McCarthy, J. and Furcht, L.T., Neurite extension by central and peripheral nervous system neurons in response to substratun-bound fibronectin and laminin, *Dev Biol*, 98 (1983) 212-220.

[152] Rudolph, A.S., Stilwell, G., Cliff, R.O., Kahn, B., Spargo, B.J., Rollwagen, F. and Monroy, R.L., Biocompatibility of lipid microcylinders: effect on cell growth and antigen presentation in culture, *Biomaterials*, 13 (1992) 1085-92.

[153] Rutkowski, G.E., Miller, C.A., Jeftinija, S. and Mallapragada, S.K., Synergistic effects of micropatterned biodegradable conduits and Schwann cells on sciatic nerve regeneration, *J Neural Eng*, 1 (2004) 151-7.

[154] Sakiyama-Elbert, S.E. and Hubbell, J.A., Development of fibrin derivatives for controlled release of heparin-binding growth factors, *J Control Release*, 65 (2000) 389-402.

[155] Saneinejad, S. and Shoichet, M.S., Patterned glass surfaces direct cell adhesion and process outgrowth of primary neurons of the central nervous system, *J Biomed Mater Res*, 42 (1998) 13-9.

[156] Satou, T., Nishida, S., Hiruma, S., Tanji, K., Takahashi, M., Fujita, S., Mizuhara, Y., Akai, F. and Hashimoto, S., A morphological study on the effects of collagen gel matrix on regeneration of severed rat sciatic nerve in silicone tubes, *Acta Pathol Jpn*, 36 (1986) 199-208.

[157] Schanen-King, C., Nel, A., Williams, L.K. and Landreth, G., Nerve growth factor stimulates the tyrosine phosphorylation of MAP2 kinase in PC12 cells, *Neuron*, 6 (1991) 915-22.

[158] Schmidt, C.E. and Leach, J.B., Neural Tissue Engineering: Strategies for repair and regeneration, *Annu. Rev. Biomed. Eng* (2003) 293-347.

[159] Schwab, M.E. and Bartholdi, D., Degeneration and regeneration of axons in the lesioned spinal cord, *Physiol Rev*, 76 (1996) 319-70.

[160] Seckel, B.R., Jones, D., Hekimian, K.J., Wang, K.K., Chakalis, D.P. and Costas, P.D., Hyaluronic acid through a new injectable nerve guide delivery system enhances peripheral nerve regeneration in the rat, *J Neurosci Res*, 40 (1995) 318-24.

[161] Sendtner, M., Holtmann, B., Kolbeck, R., Thoenen, H. and Barde, Y.A., Brain-derived neurotrophic factor prevents the death of motoneurons in newborn rats after nerve section, *Nature*, 360 (1992) 757-9.

[162] Son YJ, T.J., Thompson WJ, Schwann cells induce and guide sprouting and reinnervation of neuromuscular junction, *Trends Neurosci.*, 19 (1996) 280-85.

[163] Song, H. and Poo, M., The cell biology of neuronal navigation, *Nat Cell Biol*, 3 (2001) E81-8.

[164] Spargo, B.J., Cliff, R.O., Rollwagen, F.M. and Rudolph, A.S., Controlled release of transforming growth factor-beta from lipid-based microcylinders, *J Microencapsul*, 12 (1995) 247-54.

[165] Strassman, R.J., Letourneau, P.C. and Wessells, N.K., Elongation of axons in an agar matrix that does not support cell locomotion, *Exp Cell Res*, 81 (1973) 482-7.

[166] Strauch, B., Rodriguez, D.M., Diaz, J., Yu, H.L., Kaplan, G. and Weinstein, D.E., Autologous Schwann cells drive regeneration through a 6-cm autogenous venous nerve conduit, *J Reconstr Microsurg*, 17 (2001) 589-595.

[167] Suematsu, N., Tubulation of the peripheral nerve gap: Its history and possibility, *Microsurgery*, 10 (1989) 1-74.

[168] Sunderland, S., *Nerve Injuries and their Repair: A Critical Appraisal.*, Churchill Livingstone, New York, 1991.

[169] Suzuki, Y., Tanihara, M., Ohnishi, K., Suzuki, K., Endo, K. and Nishimura, Y., Cat peripheral nerve regeneration across 50 mm gap repaired with a novel nerve guide composed of freeze-dried alginate gel, *Neurosci Lett*, 259 (1999) 75-8.

[170] Terzis, J., Faibisoff, B. and Williams, B., The nerve gap: suture under tension vs. graft, *Plast Reconstr Surg*, 56 (1975) 166-70.

[171] Tessier-Lavigne, M. and Goodman, C.S., The molecular biology of axon guidance, *Science*, 274 (1996) 1123-33.

[172] Thanos, P.K., Okajima, S. and Terzis, J.K., Utrastructure and cellualr biology of nerve regeneration, *J. Reconstr. Microsurg.*, 14 (1998) 423-36.

[173] Thoenen, H., Barde, Y.A., Davies, A.M. and Johnson, J.E., Neurotrophic factors and neuronal death, *Ciba Found Symp*, 126 (1987) 82-95.

[174] Timple, R., Immunological studies on collagen. In G.N. Ramachandran and A.H. Reddi (Eds.), *Biochemistry of collagen*, Plenum Press, New York, 1976, pp. 319-375.

[175] Tong, X.J., Hirai, K., Shimada, H., Mizutani, Y., Izumi, T., Toda, N. and Yu, P., Sciatic nerve regeneration navigated by laminin-fibronectin double coated biodegradable collagen grafts in rats, *Brain Res*, 663 (1994) 155-62.

[176] Tresco, P., Tissue engineering strategies for nervous system repair, *Progress in Brain Research*, 128 (2000) 349-363.

[177] Uckun, F.M., Evans, W.E., Forsyth, C.J., Waddick, K.G., Ahlgren, L.T., Chelstrom, L.M., Burkhardt, A., Bolen, J. and Myers, D.E., Biotherapy of B-cell precursor leukemia by targeting genistein to CD19-associated tyrosine kinases, *Science*, 267 (1995) 886-91.

[178] Uzman, B.G., Villegas, G.M., Mouse sciatic nerve regeneration through semi-permeable tubes: a quantitative model, *J Neurosci*, 9 (1983) 325-338.

[179] Valentini, R.F. and Aebischer, P., Strategies for the engineering of peripheral nervous tissue regeneration. In L.R. Lanza RP, Chick WL (Ed.), *Principles of Tissue Engineering*, R.G.Landes Company, Austin, 1997, pp. 671-684.

[180] Valentini, R.F., Aebischer, P., Winn, S.R. and Galletti, P.M., Collagen- and laminin-containing gels impede peripheral nerve regeneration through semipermeable nerve guidance channels, *Exp Neurol*, 98 (1987) 350-6.

[181] Valentini, R.F., Sabatini, A.M., Dario, P. and Aebischer, P., Polymer electret guidance channels enhance peripheral nerve regeneration in mice, *Brain Res*, 480 (1989) 300-4.

[182] Varejao, A.S., Melo-Pinto, P., Meek, M.F., Filipe, V.M. and Bulas-Cruz, J., Methods for the experimental functional assessment of rat sciatic nerve regeneration, *Neurol Res*, 26 (2004) 186-94.

[183] Walton, R.L., Brown, R.E., Matory, W.E. Jr, Borah, G.L., Dolph, J.L., Autogenous vein graft repair of digital nerve defects in the finger: a retrospective clinical study, *Plast. Reconstr. Surg.*, 84 (1989) 944-49.

[184] Wang, G.Y., Hirai, K. and Shimada, H., The role of laminin, a component of Schwann cell basal lamina, in rat sciatic nerve regeneration within antiserum-treated nerve grafts, *Brain Res*, 570 (1992a) 116-25.

[185] Wang, G.Y., Hirai, K., Shimada, H., Taji, S. and Zhong, S.Z., Behavior of axons, Schwann cells and perineurial cells in nerve regeneration within transplanted nerve grafts: effects of anti-laminin and anti-fibronectin antisera, *Brain Res*, 583 (1992b) 216-26.

[186] Wang, K.K., Costas, P.D., Bryan, D.J., Jones, D.S. and Seckel, B.R., Inside-out vein graft promotes improved nerve regeneration in rats, *J Reconstr Microsurg*, 14 (1993) 608-618.

[187] Wang, K.K., Nemeth, I.R., Seckel, B.R., Chakalis-Haley, D.P., Swann, D.A., Kuo, J.W., Bryan, D.J. and Cetrulo, C.L., Jr., Hyaluronic acid enhances peripheral nerve regeneration in vivo, *Microsurgery*, 18 (1998) 270-5.

[188] Wang, X., Hu, W., Cao, Y., Yao, J., Wu, J. and Gu, X., Dog sciatic nerve regeneration across a 30-mm defect bridged by a chitosan/PGA artificial nerve graft, *Brain*, 128 (2005) 1897-910.

[189] Weiss, P. and Taylor, A.C., Guides for nerve regeneration across nerve gaps, *J Neurosurg*, 3 (1946) 275-82.

[190] Williams, L.R., Longo, F.M., Powell, H.C., Lundborg, G. and Varon, S., Spatial-temporal progress of peripheral nerve regeneration within a silicone chamber: parameters for a bioassay, *J Comp Neurol*, 218 (1983) 460-70.

[191] Xu XM, C.A., Guenard V, Kleitman N, Bunge MB, Bridging schwann cell transplants promote axonal regeneration from both the rostral and caudal stumps of transected adult spinal cord, *J. Neurocytol.*, 26 (1997) 1-16.

[192] Xu, X.M., Guenard, V., Kleitman, N. and Bunge, M.B., Axonal regeneration into Schwann cell-seeded guidance channels grafted into transected adult rat spinal cord, *J Comp Neurol*, 351 (1995) 145-60.

[193] Xu, X.M., Zhang, S.X., Li, H., Aebischer, P., and Bunge, M.B., Regrowth of axons into distal spinal cord through a Schwann cell seeded mini-channel implanted into hemisected adult rat spinal cord, *Eur. J. Neurosci.*, 11 (1999) 1723-1740.

[194] Yoshii, S., Oka, M., Shima, M., Taniguchi, A. and Akagi, M., Bridging a 30-mm nerve defect using collagen filaments, *J Biomed Mater Res A*, 67 (2003) 467-74.

[195] Yu, X. and Bellamkonda, R.V., Dorsal root ganglia neurite extension is inhibited by mechanical and chondroitin sulfate-rich interfaces, *J Neurosci Res*, 66 (2001) 303-10.

[196] Yu, X. and Bellamkonda, R.V., Tissue-engineered scaffolds are effective alternatives to autografts for bridging peripheral nerve gaps, *Tissue Eng*, 9 (2003) 421-30.

[197] Yu, X., Dillon, G.P. and Bellamkonda, R.B., A laminin and nerve growth factor-laden three-dimensional scaffold for enhanced neurite extension, *Tissue Eng*, 5 (1999) 291-304.

[198] Zhang, F., Blain, B., Beck, J., Zhang, J., Chen, Z., Chen, Z.W. and Leineaweaver, W.C., Autogenous venous graft with one-stage prepared Schwann cells as a conduit for repair of long segmental nerve defects, *J Reconstr Microsurg*, 18 (2002) 295-300.

[199] Zhao, Q., Dahlin, L.B., Kanje, M. and Lundborg, G., Specificity of muscle reinnervation following repair of the transected sciatic nerve. A comparative study of different repair techniques in the rat, *J Hand Surg [Br]*, 17 (1992) 257-61.

[200] Zhong, Y., Yu, X., Gilbert, R. and Bellamkonda, R.V., Stabilizing electrode-host interfaces: a tissue engineering approach, *J Rehabil Res Dev*, 38 (2001) 627-32.

[201] Zhou, F.C. and Azmitia, E.C., Laminin directs and facilitates migration and fiber growth of transplanted serotonin and norepinephrine neurons in adult brain, *Prog Brain Res*, 78 (1988) 413-26.

[202] Zuo, J., Hernandez, Y.J. and Muir, D., Chondroitin sulfate proteoglycan with neurite-inhibiting activity is up-regulated following peripheral nerve injury, *J Neurobiol*, 34 (1998) 41-54.

[203] Zuo, J., Neubauer, D., Graham, J., Krekoski, C.A., Ferguson, T.A. and Muir, D., Regeneration of axons after nerve transection repair is enhanced by degradation of chondroitin sulfate proteoglycan, *Exp Neurol*, 176 (2002) 221-8.